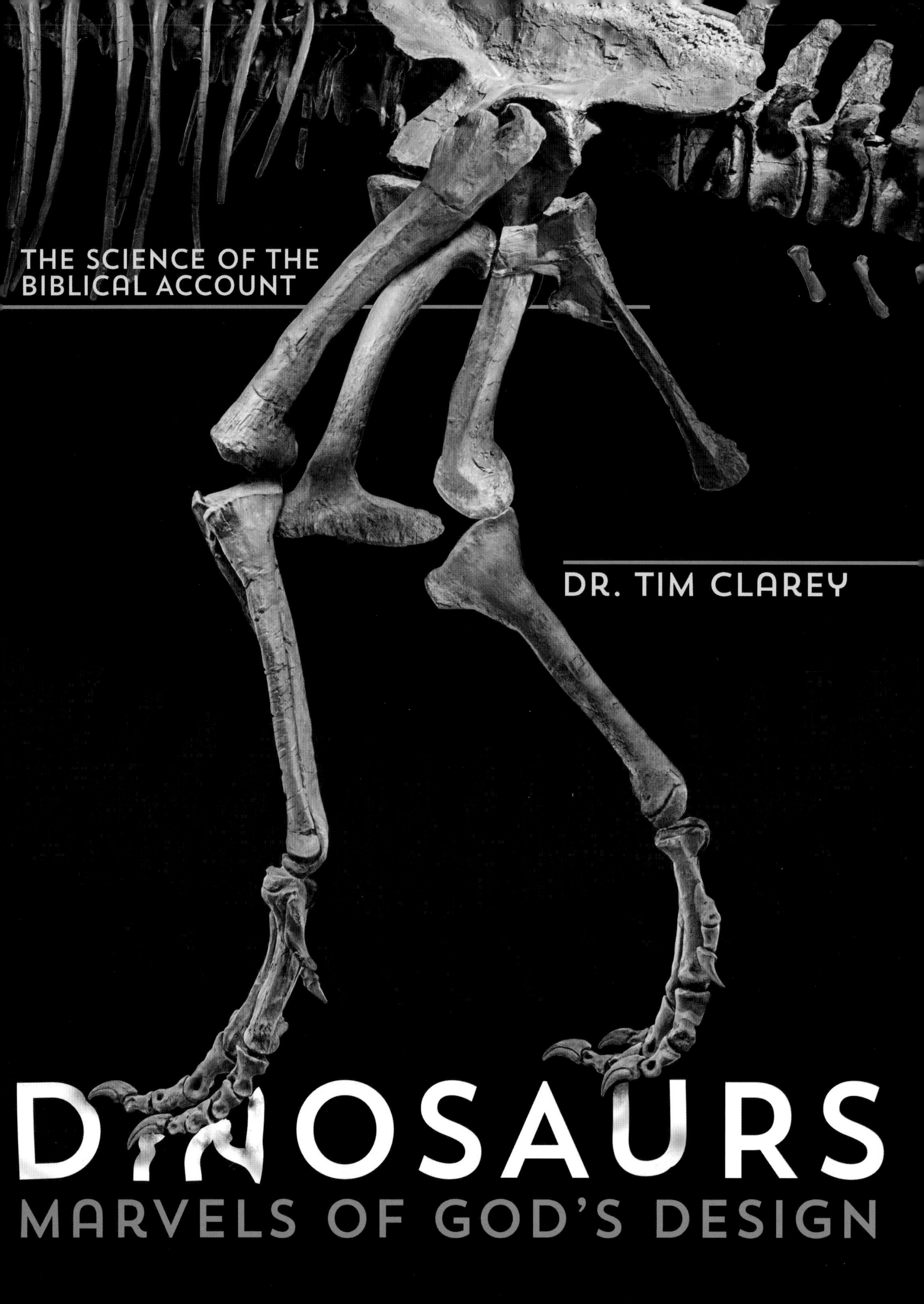
THE SCIENCE OF THE
BIBLICAL ACCOUNT
DR. TIM CLAREY
DINOSAURS
MARVELS OF GOD'S DESIGN

First printing: October 2015
Second printing: August 2018

Master Books®, P.O. Box 726, Green Forest, AR 72638

Master Books® is a division of the New Leaf Publishing Group, Inc.

ISBN: 978-0-89051-904-2
ISBN: 978-1-61458-468-1 (digital)
Library of Congress Number: 2015950629

Cover by Diana Bogardus

Unless otherwise noted, Scripture quotations are from the New King James Version of the Bible.

Please consider requesting that a copy of this volume be purchased by your local library system.

Printed in China

Please visit our website for other great titles: www.masterbooks.com

For information regarding author interviews, please contact the publicity department at (870) 438-5288.

Deinonychus antirrhopus skeleton,
Philadelphia Academy of Natural Sciences

Master Books®
A Division of New Leaf Publishing Group
com

"O Timothy! Guard what was committed to your trust,
avoiding the profane and idle babblings and
contradictions of what is falsely called knowledge—
by professing it some have strayed from the faith."
I Timothy 6:20–21

"For this they willfully forget:
that by the word of God the heavens were of old,
and the earth standing out of water and in the water,
by which the world that then existed perished, being flooded with water."
II Peter 3:5–6

PRAISE FOR *DINOSAURS: MARVELS OF GOD'S DESIGN*

If you like dinosaurs, you will love this book! Dr. Clarey and Master Books have done a marvelous job with both content and illustrations. There is exciting and captivating information within these pages for every age and every academic level. The pictures alone are worth the price! The topics flow easily from section to section, and each section provides insight to both novice and scholar alike. Dr. Clarey's research on the stages of the Genesis Flood is unique and exceedingly helpful, not only in the technical aspects of geology but in the easily understood applications of why and where the fossils are found. Good job, Dr. Clarey!

Dr. Henry M. Morris III
Chief Executive Officer
Institute for Creation Research

The topic of dinosaurs has probably been used by secularists more than another subject to confuse adults and kids concerning whether the history recorded in the Bible, especially Genesis, can really be trusted. Sadly, with the onslaught of evolutionary/millions of years teaching on dinosaurs that permeates our culture (particularly in schools), countless people have been put on a slippery slide of doubt about God's Word that often leads to unbelief. So I am thrilled to see the publication of the first truly comprehensive work on dinosaurs and how Christians should view them, a book that could best be described as a creationist encyclopedia about these fascinating creatures. It's a MUST for every family!

Ken Ham
President of Answers in Genesis
Author and lecturer on dinosaurs

Dr. Clarey has "roamed the West" for years, studying the rocks and fossils found there. He brings real expertise to this easily read treatment of a favorite subject. Prepare to be educated and enthused!

Dr. John D. Morris
President Emeritus
Institute for Creation Research

Many people seem to think that just saying the word "dinosaurs" leaves the Bible shattered and crumbling. Dr. Clarey effectively exposes this flawed thinking. Using his knowledge of geology, he shows how what we know about dinosaurs actually supports the biblical text. In fact, it supports the text very well. Dr. Clarey answers many common questions, offers some new ideas, and gives us a much needed study of dinosaurs from a creationist perspective.

Kevin Anderson, Ph.D.
Van Andel Creation Research Center

Dr. Tim Clarey's latest book Dinosaurs: Marvels of God's Design *is a masterpiece! It exhibits the latest research on the topic and all from a biblical perspective. Dr. Clarey has taught college classes on the subject of dinosaurs, and brings his expertise and years of experience to this beautifully illustrated book that even young students will appreciate.*

Dr. Jason Lisle
Author, Apologist, and Speaker
JasonLisle.com

DEDICATION

For Henry M. Morris and Duane T. Gish

For their years of service to the Kingdom and their inspiration to all that followed

"This was the Lord's doing; it is marvelous in our eyes."
Psalm 118:23

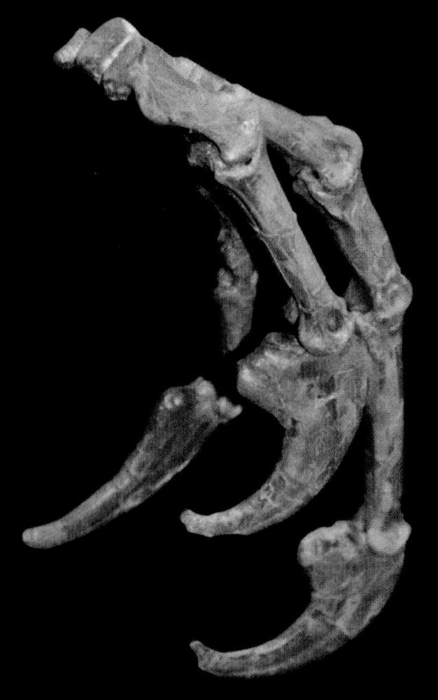

Fossil hand of a *Deinonychus* on display in the Museum of the Rockies in Bozeman, Montana. This specimen was collected in Carbon County, Montana.

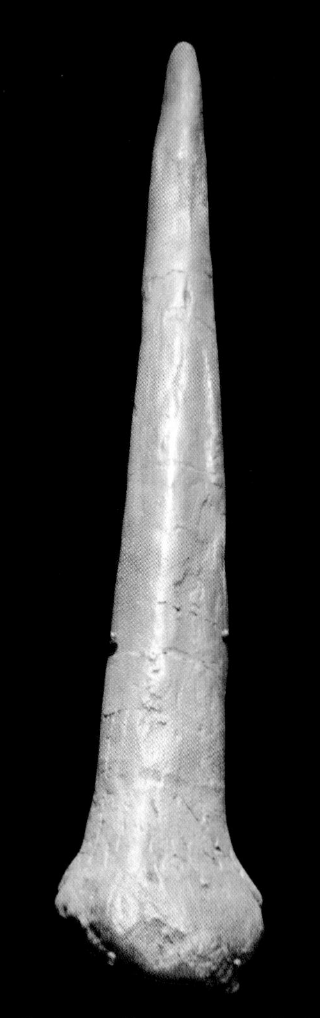

Tail spike from a *Stegosaurus* on display in the Museum of the Rockies in Bozeman, Montana.

TABLE OF CONTENTS

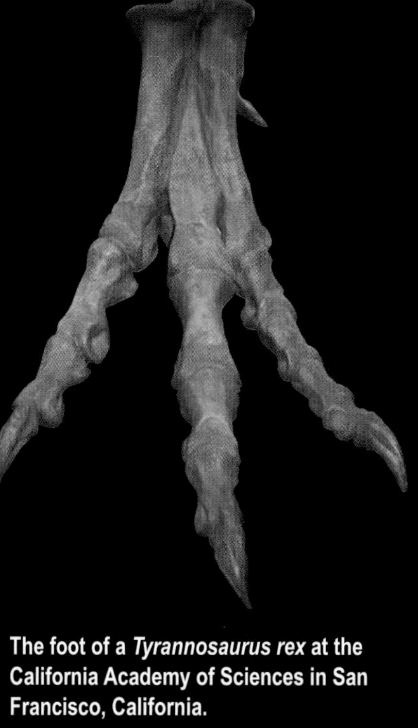

The foot of a *Tyrannosaurus rex* at the California Academy of Sciences in San Francisco, California.

Hind leg of *Brontosaurus* that was being assembled, as seen in *Popular Science Monthly* Volume 58

PREFACE

This book is vastly different from any other dinosaur book you have ever read. First of all, it explains, using sound science, that dinosaurs roamed the earth just thousands of years ago. But because of secular books and television shows, many Christians struggle to explain how dinosaurs fit in the Bible. The word "dinosaur" is not found in the biblical text. This causes some well-meaning Christians to turn to secular science for an explanation. Unfortunately, this has caused many people to lose faith in the Bible, especially the youth. This book will restore faith in the Word of God! Dinosaurs don't disprove the Bible; to the contrary, they are part of God's creative glory!

Second, this book not only explains dinosaurs in a biblical context, and using the latest science, but it covers the complete spectrum of dinosaur-related topics, from the earliest dinosaur discoveries to debate over why they went extinct. Discussions on the latest findings in dinosaur biology and behavior are also included. It also covers topics only rarely included in other creationist books, such as the significance of egg nests and footprints and how to calculate dinosaur weights and walking speeds. It concludes with a summary chapter on the "real" story of dinosaurs.

Third, this book is written as a response to critics who think creationists cannot write a truly scientific text on the subject of dinosaurs. Be assured, the text is as accurate, up-to-date, and scientifically sound as any secular book on the market. It even includes some of my own original research.

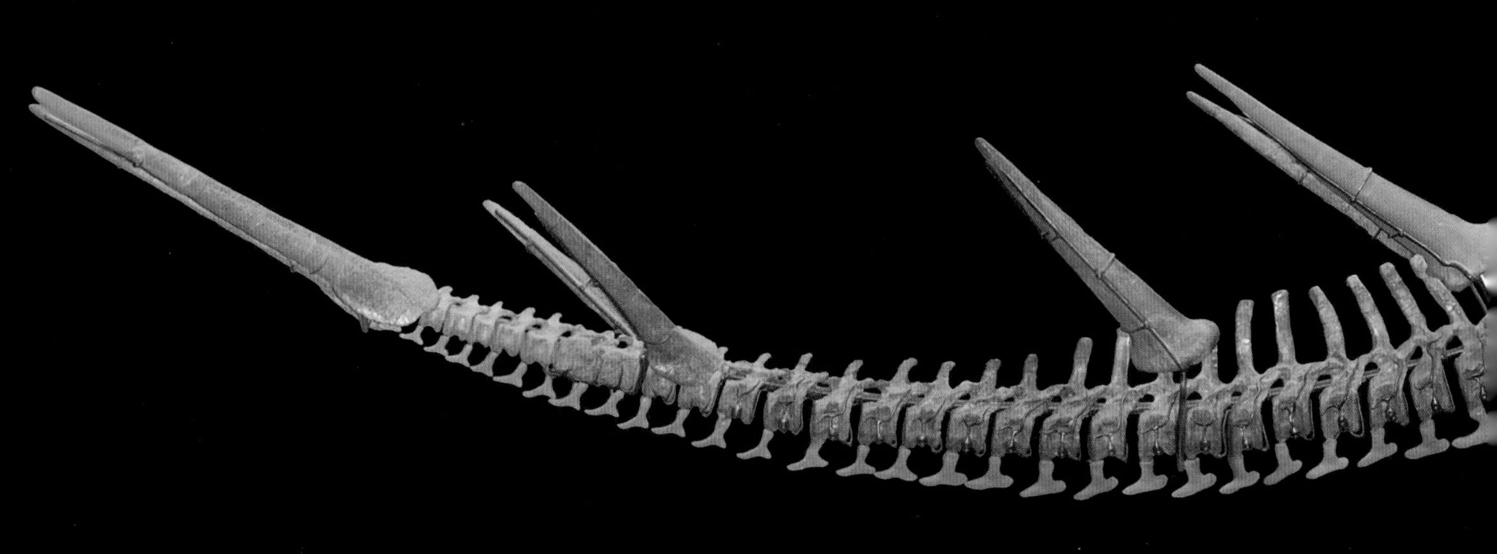

I tried to cover subjects such as dinosaur extinction and purported bird-dinosaur relationships objectively. I feel that all theories and hypotheses, no matter how popular, should be questioned, keeping in mind that popularity doesn't make it true. Facts must be distinguished from theory. I do not want readers to blindly accept anything they read or hear or see, especially in the secular media. I want you to think, taking nothing unproven for scientific fact. The facts we deal with are the Word of God, the rocks, and the fossils. How the dinosaurs went extinct and whether or not they were warm-blooded or had some type of feathered covering is still theory, and unfortunately, often based on almost no real scientific evidence at all. One thing we do know is that God created dinosaurs on day 6 of creation week (Gen. 1:24–25).

The best part of this adventure is the fun you get to have learning science in a biblical context. Dinosaurs make news headlines (and movies) almost weekly. Science is not always stagnant and boring. With each new discovery, we learn more and more about these amazing creatures called dinosaurs. I hope you, the reader, have as much fun as I do continually learning about these "terrible reptiles." Maybe this book will even inspire you to participate in a dinosaur dig like I had the privilege of doing. And yes, how to find your own dinosaur is included in here too. If you read nothing else in this book, please read chapter 14, "The Real Story of the Dinosaurs." It is a summary of the science of dinosaurs, their biblical story, and my own personal research.

Triceratops taken at the American Museum of Natural History in New York City, New York.

CHAPTER 1:

Why study about dinosaurs? Dinosaurs fascinate everybody, from young children to great-grandparents, and everyone in between. Dinosaurs have the ability to transport even the oldest person back to their childhood when they saw their first dinosaur book, read a magazine article about prehistoric creatures, or visited a museum and saw a huge skeleton. Seeing them brings out feelings of awe and wonder at their magnificence that sparks the imagination.

The author's grandson

But this is more than a science book about dinosaurs. This is a story of discovery. However, each discovery is judged on presuppositions, or a particular starting worldview, if you prefer. How factual data, like dinosaur fossils and rocks are interpreted, depends on which presupposition you start with. If you believe the earth is about 4.6 billion years old, and life appeared on earth through some evolutionary progression over millions of years, and that there was no worldwide Flood, and that dinosaurs went extinct around 65–66 million years ago, then you hold the uniformitarian worldview. Uniformitarianists (those who believe the earth has had the same processes, unchanged for eons of time) believe life somehow began from nonlife without help of a Creator, and humans are at the end of a long progression of evolution. Unfortunately, this is the dominant worldview in science today.

If, however, you believe God's Word, that man was created in the image of God just thousands of years ago, and man's sin caused the destruction of the former world, and that God sent His Son Jesus to redeem the world, you'll interpret factual data, like dinosaur fossils, much differently. This view holds that God made everything in six days and that there was a Flood that destroyed the original world just thousands of years ago. This is the presupposition used throughout this book. Dinosaurs are examined from the worldview that God's Word is true and God willingly created dinosaurs. And this worldview completely fits with the factual evidence. The study of dinosaurs, and science in general, does not conflict with the Bible; rather, they complement one another. Dinosaurs are a great way to learn about the science of God's marvelous creation. Welcome to the science and history of dinosaurs.

The official postcard of the American Museum of Natural History says this is a *Barosaurus*, and that "this unique freestanding mount is the only *Barosaurus* on view in the world"

But, you may think, don't dinosaurs disprove the Bible? Aren't dinosaurs millions of years old? Did humans ever live with dinosaurs? Questions like these abound when it comes to dinosaurs, especially from Christians. Sadly, many young people lose their faith when they study science in high school and/or college, when they are bombarded by the evolutionary worldview, even at most Christian colleges. I saw these effects firsthand when I taught college geology classes. This book leads you to the answers to these questions, and will hopefully keep people from losing their faith in God's Word.

Were dinosaurs on the ark? And how could they all fit on the ark? When did God create them? These biblical questions are also addressed in this book and answers provided on the journey of discovery.

Didn't an asteroid kill all the dinosaurs? How did dinosaurs go extinct? Didn't they just evolve into birds? They had feathers, didn't they? These are more great questions I have been asked again and again. Yes, these too are addressed in this book, and the scientific evidence to support the answers is examined in detail.

Dinosaurs were discovered by Bible-believing scientists in the early 19th century. It wasn't until the mid-late 19th century that uniformitarian thought began to dominate science, as predicted in 2 Peter 3:3–4. Scientific thought was soon taken over by the uniformitarianists, naturalists and evolutionists who claimed the world was millions of years old (now even billions) and that all life evolved. And they set out to find the "facts" to prove it, corrupting the science of dinosaurs in the process. They actively twist the truth to convince people that dinosaurs are "proof" of evolution. This book rejects that notion and shows that dinosaurs, instead, are wonders of God's creation, and their fossils are found in rapidly deposited sediment laid down in a global catastrophic Flood.

The Book of Genesis lists the order of creation week quite clearly. All the wild animals that moved along the ground, and presumably the dinosaurs, are included in the verses from Genesis 1:24–25.

In the subsequent verse (Gen. 1:26), God describes the creation of man and that He gave him (us) dominion over the dinosaurs and all the animals that had previously been created, including the birds of the air and the fish of the sea.

The Bible also indicates that there was no death (of animal kinds) before the first sin and that the dinosaurs and all other animals were created as herbivores or vegetarians (Gen. 1:29–30). It wasn't until man's sin that animals began to eat one another. Whether this happened gradually or suddenly after the first sin is unclear. Even today, paleontologists have to speculate from dinosaur teeth and jaw structure and fossil dung as to what dinosaurs ate. Teeth shape alone does not guarantee diet. Many "meat-eating" dinosaurs may have remained herbivores.

Everything was declared "good" at the end of God's creation week. Animals lived in harmony with one another and with no fear. It wasn't until after the Fall that animals began to "prey" on one another. After Adam and Eve were driven from the Garden of Eden they, and their descendants, probably lived far away from the dinosaurs, as we have found no confirmed human remains mixed in with the dinosaur fossils.

Most secular textbooks state that dinosaurs lived only from the Upper Triassic Period through the end of the Cretaceous Period, during a collection of time known as the Mesozoic Era, claimed by evolutionists to be from about 232 to 66 million years ago. However, a recent discovery in Tanzania identified dinosaur bones from rocks even deeper and deposited earlier, in units identified as Middle Triassic strata.[1] And small dinosaur footprints have been reported in Poland that go back even further in evolutionary time, to rocks from Lower Triassic strata (claimed to be 250 million years old).[2] However, creation scientists interpret these rock layers as all having been deposited during the year-long Flood that occurred just a few thousand years ago. They explain that dinosaurs are found in upper Flood rocks that were deposited later in the Flood, as they were more mobile and likely lived at higher elevations compared to many sea creatures. The energy level of the Flood also seems to have abruptly changed at the point dinosaurs and large swimming reptiles became buried in the rock strata.

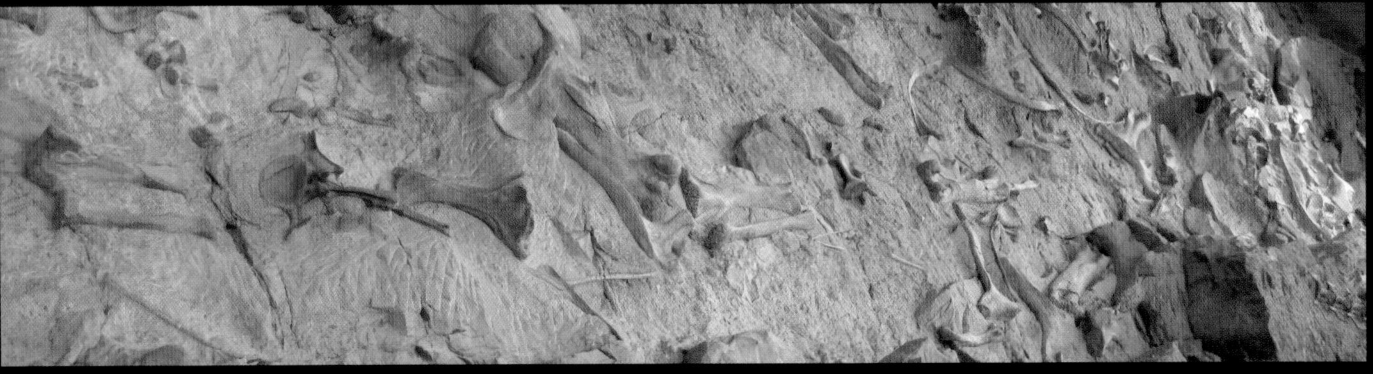

It's not just an interpretation that dinosaurs are thousands of years old. It is based on mounting factual evidence. In 2005, Mary Schweitzer published an article claiming she found preserved dinosaur soft tissue in the bone of a *T. rex* while conducting a histologic investigation (microscopic study of bone structure).[3] She was not able to extract the complete DNA sequences by any means, but has partially extracted some protein, red blood cells, and collagen. Other scientists have continued to find preserved organic remains in even dinosaur embryos. Evolutionists struggle to explain how this organic material could have survived for 70 and even 150 million years. Some nondinosaur soft tissue discoveries are even claimed to be over 500 million years old! Finding preserved organic material in dinosaur bones and other fossils fits perfectly within an age range of thousands of years ago.

Creation scientists are also finding detectable levels of carbon-14 in dinosaur bone that should have decayed to undetectable levels if the bones are truly millions of years old. These bone samples consistently show an age range of only thousands of years and not millions. In spite of these findings, evolutionists still won't admit their old dates for dinosaurs are wrong. They now think their ideas of organic preservation must be wrong. They just can't admit that dinosaurs were alive until very recently.

And God said, "Let the land produce living creatures according to their kinds: the livestock, the creatures that move along the ground, and the wild animals, each according to its kind." And it was so. God made the wild animals according to their kinds, the livestock according to their kinds, and all the creatures that move along the ground according to their kinds. And God saw that it was good

—Genesis 1:24–25 (NIV)

The "deepest" dinosaur ever discovered was found in an oil well core taken from 1.4 miles below the base of the North Sea, between Norway and Greenland (Snorre field).[4] The core contained the knucklebone of a *Plateosaurus,* a common prosauropod dinosaur found in Upper Triassic rocks of Europe. Scientists were amazed to find a terrestrial dinosaur in an area containing so many marine sediments and out in the deep ocean. Creation scientists don't find this as surprising, as many terrestrial and marine animals were often mixed together during the great Flood. It's not a mystery that some of these dinosaurs were also swept out to sea by the unimaginable violence of the Flood waters and deposited in ocean sediments.

DINOSAURS ON THE ARK

According to the Bible, dinosaurs were included on the ark. Genesis 6:20 states, "Of the birds after their kind, of animals after their kind, and of every creeping thing of the earth after its kind, two of every kind will come to you to keep them alive." All of the various kinds of dinosaurs were included in this act of God's divine providence.

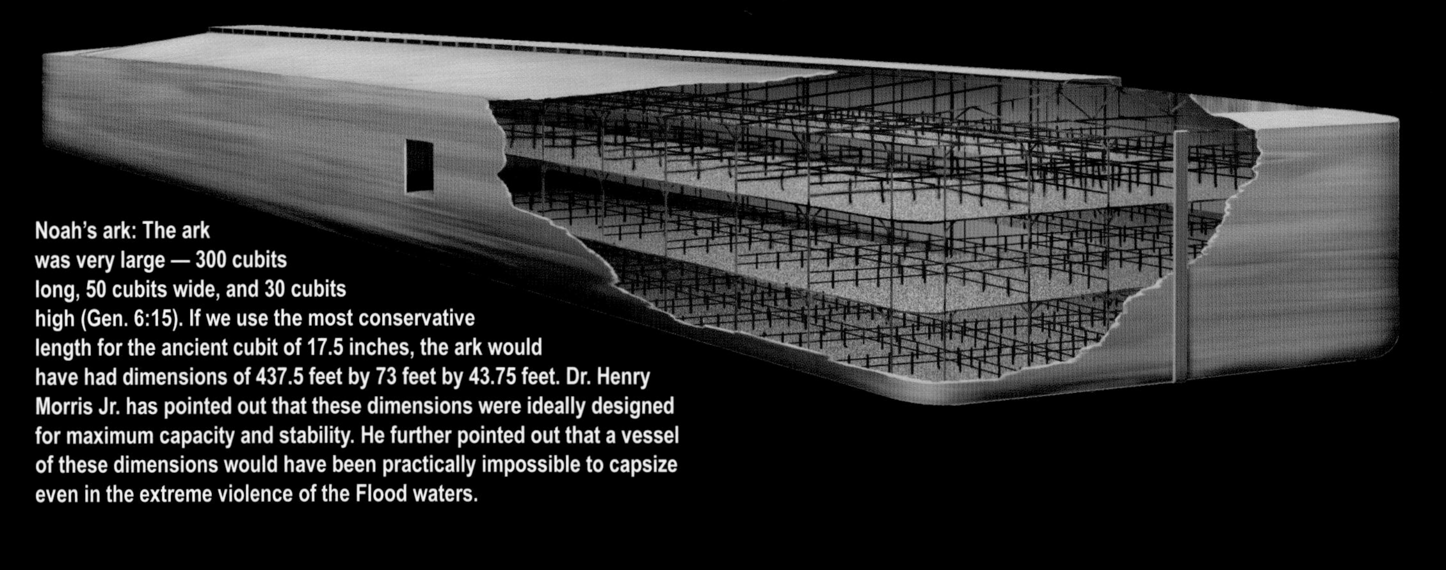

Noah's ark: The ark was very large — 300 cubits long, 50 cubits wide, and 30 cubits high (Gen. 6:15). If we use the most conservative length for the ancient cubit of 17.5 inches, the ark would have had dimensions of 437.5 feet by 73 feet by 43.75 feet. Dr. Henry Morris Jr. has pointed out that these dimensions were ideally designed for maximum capacity and stability. He further pointed out that a vessel of these dimensions would have been practically impossible to capsize even in the extreme violence of the Flood waters.

Dr. Morris calculated that the dimensions of the ark would have provided room for as many as 125,000 sheep-sized animals (155 lbs. or 70 kg.). Because many extinct and extant animals are smaller than sheep, Dr. Morris concluded that only about half the capacity of the ark would have been needed for animal storage. The other half could have been filled with water and food. God designed all animals as herbivores (Gen. 1:29) and fully capable of subsisting on plants alone. The food stored on the ark, and required to feed the dinosaurs and all animals, would have been only varieties of plants.[5]

John Woodmorappe has estimated that 139 pairs of Ornithischia (bird-hipped) dinosaurs and 195 pairs of Saurischia (lizard-hipped) dinosaurs would have been needed on the ark.[6] However, if the biblical "kind" is closer to the *family*, then this number would have been much smaller, requiring at most 60 pairs total for all dinosaurs. As you will see below, there are only 60 dinosaur families recognized by evolutionary paleontologists.

Nonetheless, many dinosaurs were much larger than a sheep, so how did they fit? In fact, the median weight of a fully grown, adult dinosaur was closer to the weight of an American bison (1,386 pounds or 630 kg), based on a study of 350 dinosaur species.[7] This may be less than most people think, considering that some adult dinosaurs were nearly half the length of a football field. But just because the median size of an adult dinosaur was fairly large, it doesn't mean there was a room problem on the ark. Most dinosaur pairs on the ark would likely have been smaller, youthful and at the onset of maturity, and not the largest and oldest dinosaurs so often seen in museums. The dinosaurs taken on the ark might have only averaged 150–220 pounds (70–100 kg), or about sheep-sized.

All dinosaurs went through a year or two when they grew very rapidly, a growth spurt if you will, similar to teenage humans (see the section on dinosaur biology).[8] Dinosaurs were possibly taken on the ark about a year prior to this growth spurt, thereby needing less to eat over the course of the year-long Flood. After the Flood, upon their release to the land, the dinosaurs would have experienced their growth spurt, rapidly maturing to adult size, and presumably sexual maturity, and therefore, able to fulfill God's command to repopulate and fill the earth (Gen. 9:7). In order to fulfill this command, the so-called meat-eating dinosaurs (theropods) probably ate only plants after the Flood, at least for some time, before returning to their meat-eating ways.

Animals of all kinds came to the ark on God's command, including about 60 kinds of dinosaurs.

ICR'S JUVENILE DINOSAUR "EDDIE"

In 2008, the Institute for Creation Research in Dallas, Texas, acquired "Eddie," a rare juvenile *Edmontosaurus* (duck-billed hadrosaur). This "little" dinosaur is about 10 feet long and 5.5 feet tall. He was discovered in 1990 in Montana's Upper Cretaceous Two Medicine Formation, which is famous for its dinosaur discoveries, including thousands of bones from the *Maiasaura* — a type of large, duck-billed dinosaur.

Because juvenile dinosaurs are quite rare, little is known about the growth patterns of most dinosaurs. However, the Two Medicine Formation provides specimens from many stages of growth of the *Maiasaura* — enough to allow scientists to plot their growth history. *Edmontosaurus* and *Maiasaura* are very similar, falling into the same subfamily of Hadrosauridae, which likely places these two genera in the same biblical "kind."

Using the growth history of the *Maiasaura*, we can estimate Eddie's age from his size. Because we didn't want to damage the specimen by cutting a leg bone in half to count growth rings (similar to tree rings), we used an equation to estimate Eddie's weight based on the circumference of his femur (see section on "How Heavy Is That Dinosaur" in chapter 10). Scientists have measured the femur (the main weight-bearing leg bone of two-legged animals) of various living animals at the midpoint where the bones are the thinnest (to determine the minimum structural support necessary). Using these numbers, researchers have found a best-fit equation for weight versus femur circumference for all two-legged animals, including dinosaurs.

The beauty of the circumference method is its simplicity — all we need for a weight estimate of a given fossil specimen is a leg bone, and leg bone fossils are often well-preserved. So how does Eddie weigh in? The circumference of the thinnest point of his femur is 167 mm. By using the bipedal equation given in chapter 10, we arrive at a weight estimate of 412 pounds, or 187 kilograms (1 kg = 2.2 lb).

Given this weight, how old was Eddie when he died? If we use the published growth curve for the *Maiasaura*,[9] we see that duck-billed dinosaurs grew slowly for the first four years and then hit a rapid growth spurt between ages five and six. During that time, hadrosaurs could have been gaining as much as 2,292 pounds (1,042 kg) per year in body weight. By the time they reached six or seven years old they would have been nearly adult size at over 3,300 pounds (1,500 kg) and presumably becoming sexually mature. An adult *Edmontosaurus* could reach 37 feet in length and stand about 18 feet tall.

***Brachiosaurus* femur, or rear thigh bone, and geologist Metrinah Ruzvido in Zimbabwe.**

Eddie's 412-pound weight places him in the four-year-old range, just before the onset of the growth spurt. A hadrosaur dinosaur of this age and size would have been a perfect candidate for Noah's ark. Unfortunately, Eddie wasn't on the ark — he died by rapid and catastrophic burial in sediment during the Great Flood, only to be found by paleontologists later and put on display as a witness to this judgment event.

God may have placed a similar pair of four-year-old hadrosaurs on the ark, knowing these dinosaurs were the perfect age and size for the journey. They would not require much room or much to eat during the year-long Flood. Upon leaving the ark at age five, however, they would have required a lot of food as they hit their growth spurt. In addition, these dinosaurs would likely have become sexually mature soon after leaving the ark, able to quickly fulfill God's command to multiply upon the earth (Genesis 8:17).

WHAT IS A DINOSAUR?

People of many cultures had found dinosaur bones for many centuries before dinosaurs were recognized. They explained their finds as dragon bones, giant human bones, or other "beasts" of the field. The name "Dinosauria" was first used by Sir Richard Owen in 1841, in an address to the British Association for the Advancement of Science, subsequently publishing the term in 1842. He was the first to recognize that dinosaurs (from the Greek, meaning "fearfully great reptiles") were a distinct group of reptiles, much different from today's lizards. Owen defined dinosaurs as reptiles that walked erect, having a posture similar to elephants and rhinos, where the legs come straight down from their bodies. He realized this after examining their hip structure and the holes in the hip sockets.

However, there were some early paleontologists who still thought dinosaurs walked in the sprawling, belly-dragging style of alligators and crocodiles. Around 1910, several German paleontologists, including Gustav Tornier, backed the sprawling leg style of lizards for their reconstructions of dinosaurs. The matter was eventually settled after examination of the large rib cages of dinosaurs that necessitated long, straight legs to elevate the body frame above the ground. Other paleontologists pointed out the sprawling posture would have forced the legs to be disjointed. Footprint data have also confirmed the upright posture. Thousands of dinosaur footprints show a pace angulation that places the feet close together (no sprawl) and no belly-dragging marks (with one possible exception where the dinosaurs were mired up to their hips in mud). There is no scientific dispute on dinosaur posture today.

Sir Richard Owen was the best known and most authoritative comparative anatomist and reptile expert in the mid-19th century. He picked up the scientific baton (so to speak) right where France's Baron Georges Cuvier left off as the "go-to guy" in animal anatomy. Cuvier was a wonderful scientist and broke significant barriers in his studies. He was the first to point out that some animals had gone extinct, startling the scientific world, in an announcement he made in 1796.

Amazingly, up until that time, no one had thought anything could go extinct. It was wrongly believed that God wouldn't allow it. But, the sin of man and the Curse

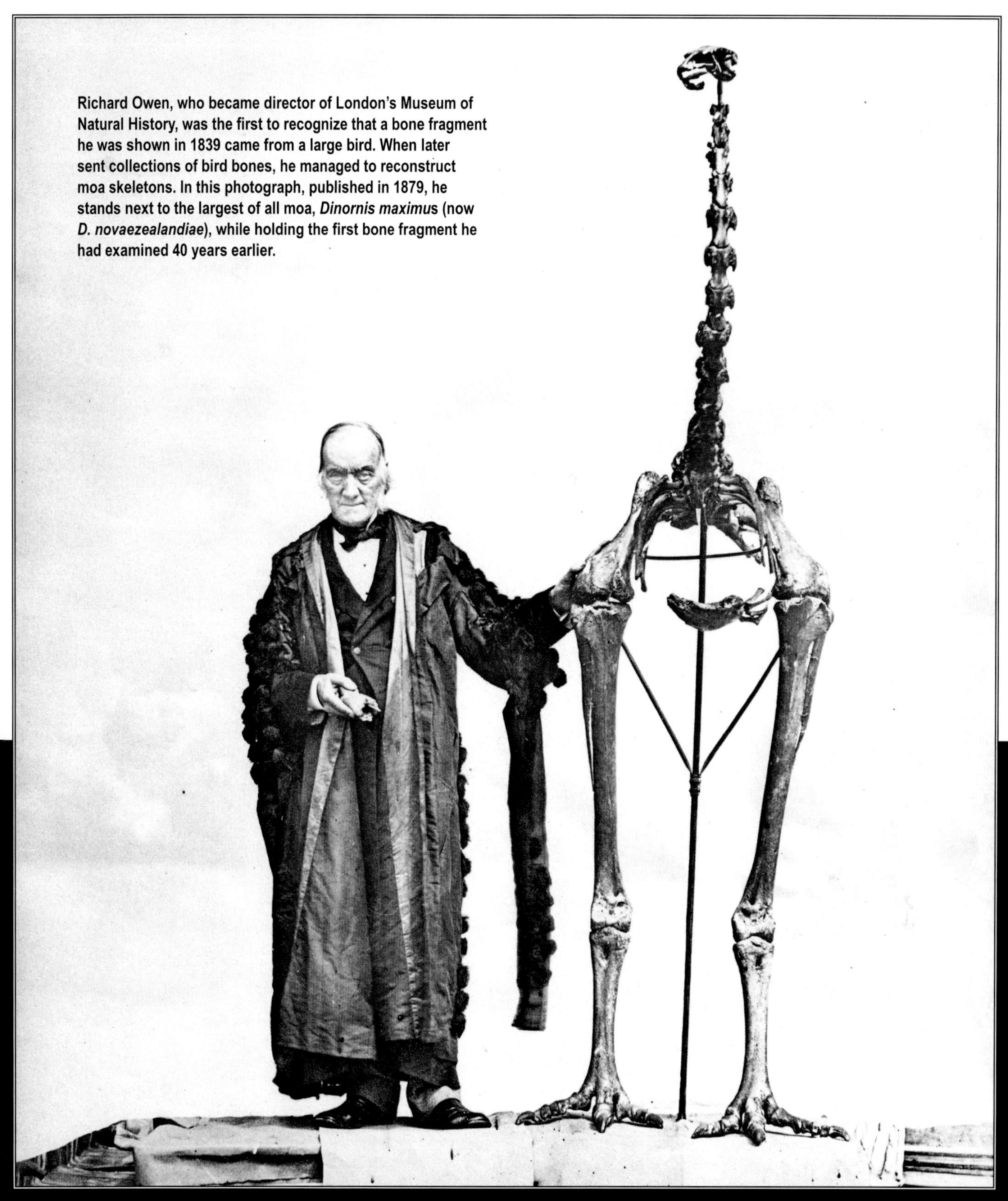

Richard Owen, who became director of London's Museum of Natural History, was the first to recognize that a bone fragment he was shown in 1839 came from a large bird. When later sent collections of bird bones, he managed to reconstruct moa skeletons. In this photograph, published in 1879, he stands next to the largest of all moa, *Dinornis maximus* (now *D. novaezealandiae*), while holding the first bone fragment he had examined 40 years earlier.

had caused many species to disappear after the Flood, and Sir Richard Owen had identified an entire group of now-extinct reptiles. Later in his life, Owen was also an outspoken anti-evolutionist and was critical of Charles Darwin's theory, arguing extensively against evolution. This is no small point. Here was the 19th century's most knowledgeable anatomist and animal expert, and he didn't see any evidence to support evolution.

Another British paleontologist, Harry Seeley, later divided the dinosaurs into two categories based on more subtle hip differences. He called one group the Ornithischia, or bird-hipped style, and the other the Saurischia, or lizard-hipped style. All dinosaurs can be classified into one hip style or the other. Although the pelvic bones of the Saurischia resemble a lizard, these dinosaurs still had holes in their hip sockets for walking upright.

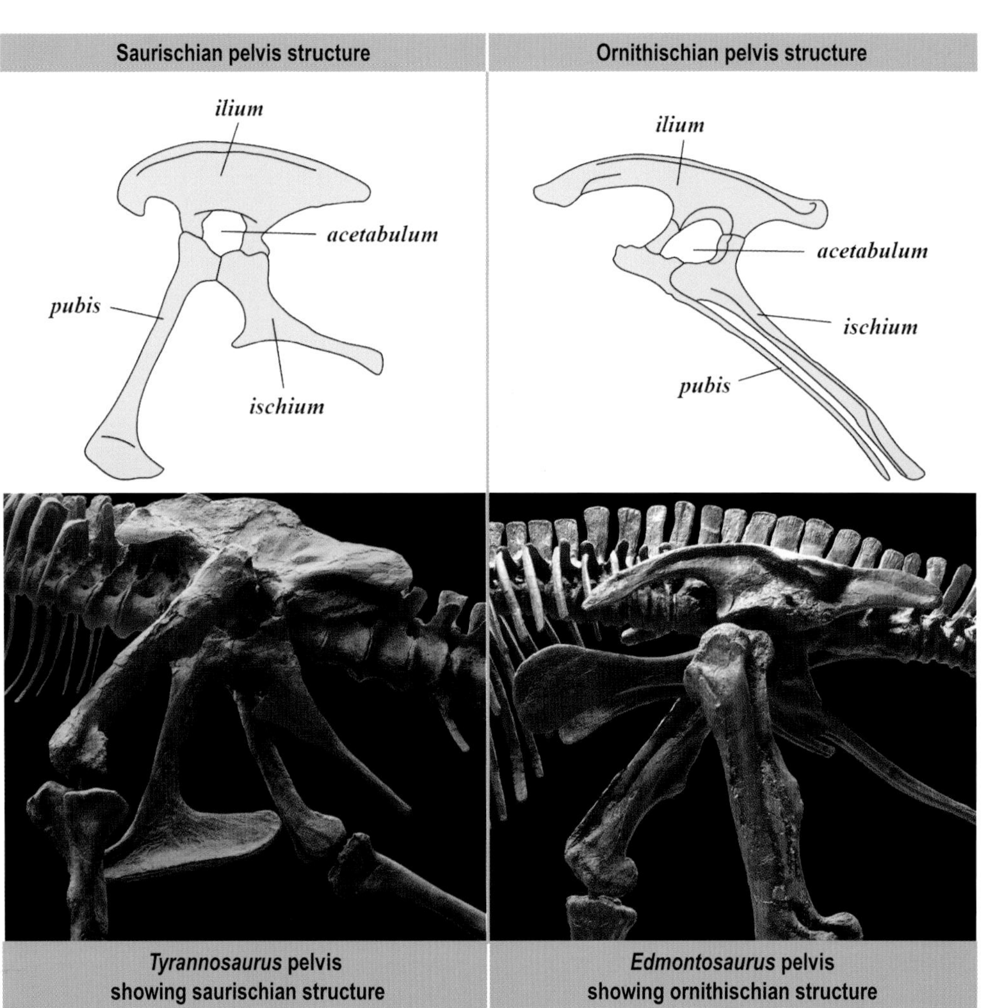

Tyrannosaurus pelvis showing saurischian structure

Edmontosaurus pelvis showing ornithischian structure

Dinosaurs did not have wings, flippers, or fins. Those are, instead, flying reptiles and swimming reptiles, respectively. Even though they are found in the same rock strata as dinosaurs, they were not dinosaurs as defined by Owen. Dinosaurs walked, and occasionally ran, on land. Some had "sails" on their backs like *Spinosaurus*, but the one famous sail-backed reptile was not a dinosaur at all. It was, instead, a lizard-like animal called *Dimetrodon*. Many fossil reptiles, like *Dimetrodon* (which grew to lengths of nearly 10 feet), are similar to lizards of today, as they have sprawling hips and legs that extend outward from their body.

All of the theropod dinosaurs (many of which became meat-eaters after the Fall) and the sauropodomorph dinosaurs (the long-necked herbivores) have similar, lizard-style hips (Saurischia). All other types of dinosaurs had birdstyle hips (Ornithischia). It is rather ironic that the so-called, most bird-like dinosaurs (theropods) had lizard-style hips. This should be a "red warning flag" to the evolutionists who want to make dinosaurs birds and vice versa.

Swimming reptiles like Mosasaurs, Plesiosaurs, and Ichthyosaurs lived exclusively in the marine realm and did not walk on land, nor did they have the hip structure of dinosaurs. Origins of the Ichthyosaurs, and all marine reptiles for that matter, remain a problem for evolutionists, as they appear suddenly in the rock record perfectly designed for aquatic life. It appears that these reptiles may not have even laid eggs on land as the sea turtles do today. One spectacular fossil has been found that shows an Ichthyosaur giving live birth in the water, just like dolphins and porpoises.

Pterosaurs and other flying reptiles are also not truly dinosaurs as defined by Owen. They are quite unique, however, and were designed to fly completely differently than birds or bats. Birds fly primarily with feathers, and bats support their wings with several long finger bones. Flying reptiles, by contrast, have a single finger made of extended bones, starting with the fourth metacarpal to which is attached four elongate phalanges. This one long finger makes the base support for the entire wing membrane. The remaining three fingers serve as short claws on the front of the wings.

DINOSAURS IN HUMAN HISTORY

Because the term "Dinosauria" was not invented until the mid-19th century, we cannot expect the Bible to contain the word "dinosaur," since the Bible was translated into English centuries earlier. But there are descriptions of creatures in ancient texts, including the Bible, that strongly suggest a connection to known kinds of dinosaurs. The Book of Job, probably the oldest book of the Bible, reported:

> Look at Behemoth, which I made along with you and which feeds on grass like an ox. What strength it has in its loins, what power in the muscles of its belly! Its tail sways like a cedar; the sinews of its thighs are close-knit. Its bones are tubes of bronze, its limbs like rods of iron.
>
> —Job 40:15–18 (NIV)

This ancient passage seems to describe a huge, long-necked-type animal (called a sauropod) that ate grass. Until recently, most secular paleontologists believed that grasses did not exist at the time of dinosaurs. However, new discoveries have shown sauropods did eat grass. Paleontologists have found evidence of grass in fossilized dinosaur dung (coprolites) from Upper Cretaceous rocks of India. Even the statement "tubes of bronze" makes one wonder if the author had "inside" knowledge about dinosaur bones long before they were rediscovered, as sauropod vertebrae are hollowed out with structures called pleurocoels along the sides of the vertebrae. These openings could be accurately described as air-filled tubes as we read in the Book of Job. Sauropods also had thick, massive leg bones as strong as "iron." It appears this passage was an extremely sophisticated description of a sauropod dinosaur right down to the tail being like a cedar tree.

"Dinosaur" rock art along the eastern shore of Lake Superior, Ontario, Canada.

Dragons have also been reported by Alexander the Great (4th century B.C.), John of Damascus (c. A.D. 675–749) and Marco Polo (13th century A.D.). Ancient historians, including Josephus (1st century A.D.) and Herodotus (4th century B.C.) also mentioned dragons or winged serpents. Petroglyphs (carvings into stone) and pictographs (paintings on stone) occasionally show creatures that strongly resemble dinosaurs, created by native tribes, especially in the southwestern United States region. Unfortunately, most of these images are interpreted by secular archaeologists as some type of "mythical beast" and not the dinosaur they resemble.

The fragmented pieces to the story of Gilgamesh were first discovered in 1844 by a British man named Austen Henry Layard while excavating the ruins of the ancient library of Nineveh. Over 25,000 fragmented pieces were sent to the British Museum in London. George Smith, a curator, began to translate the fragments in 1872, noticing that the pieces held the story of a "Babylonian Noah" and a Chaldean Flood account, similar to the Bible. Because the earliest pieces to the Gilgamesh story date to around 2100 B.C., predating the written Book of Genesis, many secular scholars have claimed that the Hebrew text and story of the Flood is based on these earlier tales, and that they merely copied and wrote down their own version. To the contrary, creation scholars believe that the account recorded in the Book of Genesis supersedes all other Flood accounts. The saga of Gilgamesh simply records a nonscriptural account of the worldwide calamity that influenced every other culture. Every one of the more than 200 "Flood" stories from all over the globe is a bit different from the biblical account. Most of these stories were passed from generation to generation, verbally, creating slight differences here and there during the transcription. However, the many repetitive similarities in all the stories cannot be accounted for by mere chance. The universal Flood event left a lasting impression on every culture of the world. And the true, inspired account of this event is recorded in the Book of Genesis.

Tablet XI of the Epic of Gilgamesh, also called the Flood Tablet, is written in Akkadian

One of the strongest lines of evidence that humans and dinosaurs interacted after the Flood was found on Ta Prohm Temple at Angkor, Siem Reap Province, Cambodia. This 12th–13th century carving depicts a beautiful rendition of a *Stegosaurus*-like animal, created about 800 years before the discovery of any fossil bones from *Stegosaurus*. For many years, paleontologists debated how the plates of a *Stegosaurus* were arranged. Did they stick up along the back or lie down flat? Maybe they should have examined ancient art for a clue. The carving of this animal shows the plates sticking up along the back in a very modern, museum rendition.

***Stegosaur*-like carving on Ta Prohm temple in Cambodia**

The Narmer Palette, an ancient Egyptian artifact carved into siltstone, depicts human wranglers controlling two creatures that resemble long-necked dinosaurs. The animals on the plaque have legs that extend straight down from their bodies, and serpentine necks, two features found in sauropod dinosaurs. This artifact was most likely created within a few centuries following the Flood, and also suggests that dinosaur-like animals coexisted with humans.

The term "dragon" appears over 20 times in the Bible, translated into English (in the King James Version, which preceded evolutionary influences on translation vocabulary) from the related Hebrew words *tannin* and *tannim*. Psalm 74:13 refers to sea dragons, like swimming reptiles, possibly a plesiosaur or mosasaur; Isaiah 30:6 refers to flying serpents, like flying reptiles, possibly a pterosaur; and Isaiah 34:13 refers to land dragons, possibly a dinosaur.

So there are many, many accounts of humans and dinosaurs over the last few thousand years. Most are, of course, discounted by mainstream scientists, but the evidence does lend strong support for human and dinosaur interaction for many centuries after the Flood.

Both sides of the Narmer Palette with depictions that look like Egyptians holding sauropod dinosaurs using ropes.

HOW MANY DINOSAUR KINDS?

Our modern classification system lists well over 1,000 species of dinosaurs, with 20 to 30 new species named each year. Unfortunately, there seems to be an attitude, among paleontologists, to hastily name a new species before someone else beats them to it. Most secular scientists are "splitters" not "lumpers." Splitters name a new species on the basis of the slightest difference in anatomy. This way, they are able to get a journal publication from their discovery. More publications lead to more grants and more academic promotions, and possibly more fame. So there's an incentive to name as many new dinosaurs as possible.

This attitude has led to many mistakes and numerous synonyms. In other words, the same dinosaur species has often been named more than once (i.e., *Brontosaurus* and *Apatosaurus*). This rather disturbing trend started back in the latter half of the 19th century with O.C. March and E.D. Cope competing to name the most dinosaurs. Marsh and Cope named many species on just a few bones and rarely published drawings for comparison. Between 1870 and 1899, Marsh named 80 different dinosaur species, but only 29 percent were subsequently found to be new or distinct. The other 71 percent he named were synonyms or repeats that he or someone else had already named. He eventually ended up with only 23 distinct species to his credit. Cope had a rather dismal dinosaur-naming success rate of only 14 percent but still managed to name nine new species that withstood later review. Dr. Michael Benton estimated that the overall success rate in naming a new dinosaur species has been about 50 percent. John (Jack) R. Horner has stated that he believes as many as 40 percent of known dinosaur species may be repeats of the same species, just in different stages of growth and development. A lot of this is due to the incomplete nature of most dinosaur discoveries.

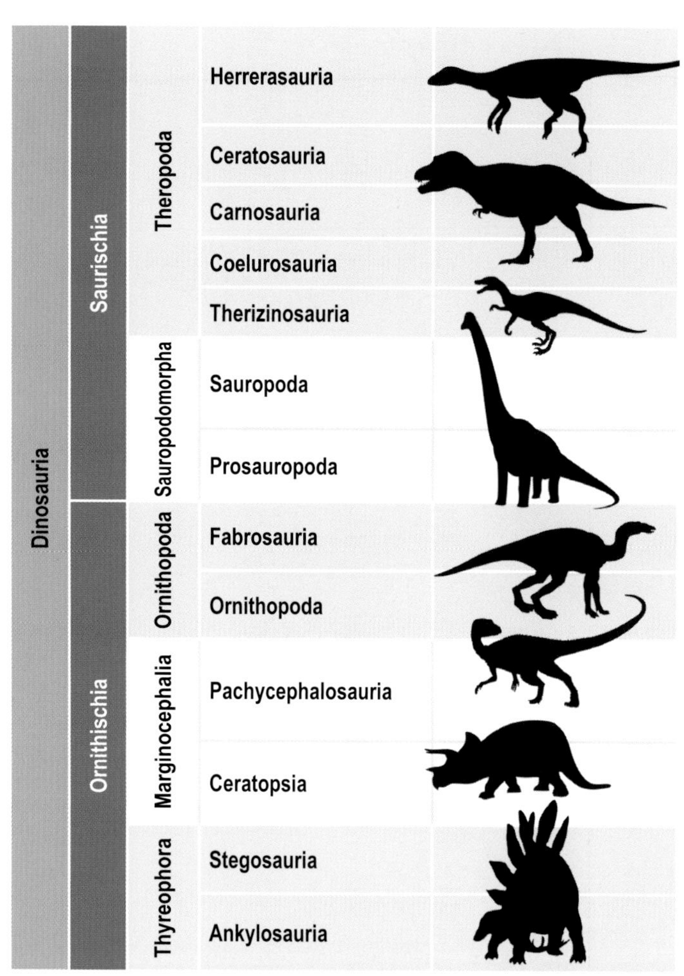

One approach to classifying the Dinosauria into Dinosaur Orders (2), Suborders (5), and Infraorders (13).

John Ostrom and Peter Wellnhofer proposed that the 16 species of the genus *Triceratops* (a ceratopsian) could be reduced to just a single species. They believed that all variations observed within each of the the 16 species are no different from the observed variations seen in modern bovids (i.e., cattle). There is even convincing evidence that *Triceratops* and *Torosaurus* may be the same genus.[10] After studying over 50 specimens of these dinosaurs, it was concluded that *Triceratops* may be merely an older juvenile *Torosaurus*, possibly eliminating the need for the genus *Triceratops* altogether. In the future, don't be surprised if one of your favorite dinosaurs has to change its name.

Determining which modern species belong within the same biblical "kind" is an exciting area of research. Creation biologists are actively working out the limits to variation within a "kind," to better understand where it falls on our modern taxonomic system. Dinosaurs are presently divided into orders, suborders, infraorders, and families. There are only five accepted suborders, 13 infraorders, and 60 families of dinosaurs. A family seems a reasonable approximation of a "kind."

The definition of a species is also difficult to quantify. It becomes even more difficult with extinct species, as we do not get to observe behavior. Biologist Ernst Mayr defined the term species as "groups of actually or potentially interbreeding natural populations, which are reproductively isolated from other groups."[11] Unfortunately, paleontologists cannot observe the reproductive habits of dinosaurs to sort out species, and therefore, any small variation is usually called new species.

Creationist scientists point out that there are a lot less "kinds" of dinosaurs than those named as a species. The Bible teaches that God made discrete kinds of organisms during the creation week that cannot interbreed. There are natural limits and boundaries to these kinds that have been set by God. Amazingly, God has placed in each kind the ability to diversify into many "breeds" or even "species," which nonetheless remain the same basic "kind." For example, dogs have diversified into retrievers, beagles, dachshunds, and so on. But they remain dogs. There is no biblical reason to assume a "kind" is the same as a "species" or that it lines up perfectly with any particular level of our modern taxonomic system.

DINOSAUR ANCESTORS?

One of the biggest problems with the evolutionary story of dinosaurs is the lack of fossil evidence to support it. There are just no undisputed ancestors for any kind of dinosaur to back the story described in secular textbooks. Every type of dinosaur appears suddenly in the rock strata, fully formed. There are no intermediate styles and no transitional varieties between any of the dinosaur groups. The only links are imaginary, not factual.

Spinosaurus specimen. Makuhari Messe in Chiba, Japan.

Evolutionists cannot even agree on the type of ancestor they seek for the "first" dinosaur. Was it a crocodile-like animal? Was it warm-blooded? What did it look like? Without any fossils to back their claims, secular scientists are merely left with speculation and guessing.

One paleontologist stated, "The demise of *T. rex* and most dinosaurs some 65 mya may grab headlines. But paleontologists are equally concerned with puzzling out how these mighty beasts got their start. Who were their ancestors?"[12]

Creation scientists realize there are no dinosaur ancestors to be found. The dinosaur fossils we find today were buried in the Flood, layer by layer, group by group, without transitional types, as they never existed. God created all the groups of dinosaurs on day 6 of creation week, fully formed with lots of diversity and variation. The different groups of dinosaurs we find in different rock strata likely reflect the various ecological communities in which the dinosaurs lived. The order of deposition during the Flood simply preserves these differences.

Washington, District of Columbia. A woman touches a dinosaur skull at a museum.

CHAPTER 2:

Carolus Linnaeus
1707-1778

Dinosaur paleontologists use the same name classification system that is used in biology. The Linnaean hierarchy is used for all extinct and extant organisms. This robust system was established by the Swedish naturalist and Creation scientist Carolus Linnaeus in the 18th century. Linnaeus developed this system in an attempt to delineate the "kinds" referred to in the Book of Genesis. He believed that there could be limited variation within the "kind" but that each kind was incapable of changing from one type to another.

Carl Linnaeus lived from 1707–1778. Born in Sweden, he traveled to the Netherlands and earned his doctorate in medicine in less than two weeks at the age of 28. He returned to Sweden and published his two-volume, 1,200-page book, *Species Plantarum,* that established the modern botanical nomenclature system in 1753. He was knighted by the king of Sweden that same year and ennobled in 1757, taking the name Carl von Linne.

The Linnaean hierarchy now begins with three domains as the highest level of classification. Domains are subdivided into kingdoms, and then further subdivided into the smaller groupings of phylum, class, order, family, genus, and finally, species, respectively. Latinization of the formal scientific name is mandatory in this system as is the italicization of all genus and species names. Also, each species is properly designated with both a genus and species name. The genus is capitalized and the species name is not. Although the genus can be written alone, the species name is never written without the genus or an abbreviation of the genus. In other words, it is improper to use *rex* alone. It must be called a *Tyrannosaurus rex* or *T. rex.* It is also improper to use hyphens in the names as is so popular with T-rex.

It is fairly easy to name a new dinosaur. One merely looks down a list of Greek and/or Latin root words and finds one that describes a particular feature about the new specimen for the genus name. "Tyranno" means tyrant, for example. Then simply add a second Greek or Latin root word such as "saurus," which means lizard or reptile. Occasionally, paleontologists may even add three root names together when naming a new genus.

The species name can honor the discoverer or someone affiliated with the discovery, but it must be Latinized too. Paleontologists digging in Madagascar found a new dinosaur they named *Masiakasaurus knopfleri,* which means "vicious lizard knopfler." Mark Knopfler is a singer/songwriter for the band Dire Straits. The scientists were listening to Dire Straits as they dug and made the discovery.

CAROLI LINNÆI REGNUM ANIMALE.

I. QUADRUPEDIA. II. AVES. III. AMPHIBIA. IV. PISCES. V. INSECTA. VI. VERMES.

PARADOXA

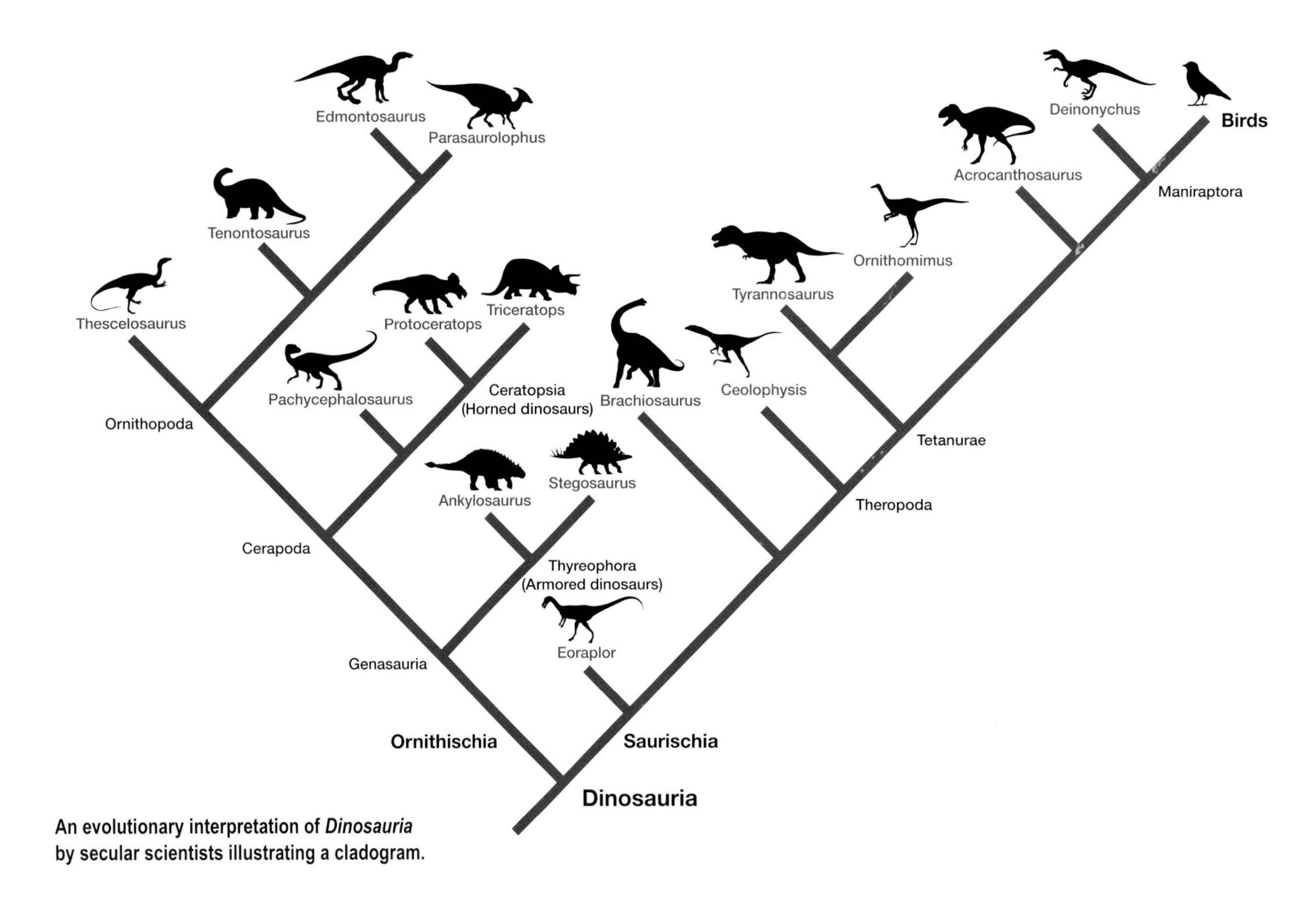

An evolutionary interpretation of *Dinosauria* by secular scientists illustrating a cladogram.

CLADISTICS: PALEONTOLOGY'S NEW "TREE OF LIFE"

Recently, many paleontologists have proposed a change in the way we classify organisms. This new system is called cladistics. It was first proposed by German entomologist Willi Hennig. Cladistics describes the pattern of life with no relation to ancestor/descendant identification. Cladograms, as the diagrams are called, have no time axis. In addition, cladistics classifies extinct and extant organisms equally, looking only for morphological patterns. In this way, unknown or missing transitional fossils are conveniently no longer problems for the paleontologist. They just become missing branches on their diagrams, and are designated as "unknown ancestors." Creationists maintain that these unknown ancestor fossils will never be found, as they never really existed.

Cladists use "evolutionary novelties" to form branches or nodes in their diagrams. This terminology is, in itself, loaded with evolutionary assumptions. An evolutionary novelty is defined as an "inherited" change from a previous pattern or structure that makes the organism unique. Examples of evolutionary novelties for mammals include warm-bloodedness, hair, and live birth. Paleontologists can still "test" the closeness of the fossils on the basis of common evolutionary novelties. The better the patterns match, the closer the location on the diagram. Whether or not we ever find these "unknown ancestors" is not the point. They become virtual "place-holders" on the cladograms.

The biggest problem with cladistics is the arbitrary choice of what constitutes an evolutionary novelty. This is particularly true of extinct organisms where we only have morphological features to compare. We cannot use warm-bloodedness as an evolutionary novelty when classifying dinosaurs and other reptiles. We may never know which, if any, dinosaurs were warm-blooded. The choice of which evolutionary novelty to use or not to use is still based on someone's opinion. This type of "truth" is unscientific and untestable. Evolutionists can assure the answer they want by manipulation of the evolutionary novelties. This is how birds have become reclassified as dinosaurs in the minds of so many evolutionists.

Birds and dinosaurs are placed close together on most secular cladograms. Evolutionists have argued that over 100 evolutionary novelties support this interpretation. However, many of these novelties were hand selected to make birds appear to be closer to dinosaurs than they actually are. Critics have pointed out that many of these 100 features overlap and are co-correlated, meaning just one feature may be represented by 15 different nodes, padding their results, and obtaining the fit they desire.[1]

Creationists, and some paleontologists, are reacting critically to the overemphasis on cladistics at the exclusion of other data. They warn that a strict application of cladistics often ignores stratigraphy, embryology, and other geologic data sets that may contradict their interpretations.

Traditional family trees force the scientist to assume common descent as diagrams branch out and upward, forming a tree. These diagrams force ancestor and descendant relationships to everything on the diagram. Creationists are critical of these diagrams also because they rely on transitional forms between groups that have never been found. Evolutionists constructing these diagrams also assume similarity equates to common ancestry instead of common design.

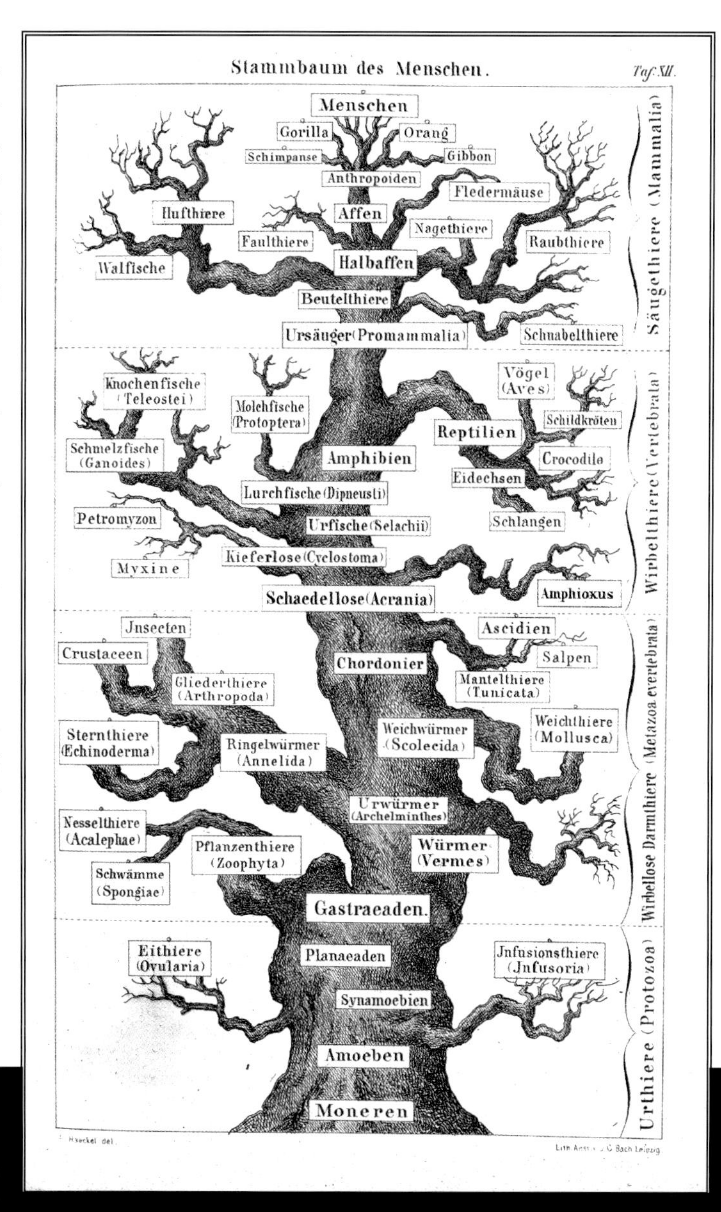

Older "family tree" style evolutionary diagram.

DISTRIBUTION OF DINOSAURS

How did dinosaurs get from continent to continent? Was the pre-Flood land mass really that different? Let's look at the latest science to try and answer these questions. It is true that paleontologists have found dinosaurs on every continent, including Antarctica. Many dinosaurs have even been found on several continents now separated by vast oceans. The best geological evidence demonstrates that the continents were not always where they are today in relation to one another. In the pre-Flood world, most of the continents were connected in one, great land mass. During the year-long Flood, the continents separated rapidly, at rates of several feet per second, into their current positions. Creation geologists have termed this *catastrophic plate tectonics*, and believe that this breakup occurred while Noah and the animals were protected on the ark.

Plate tectonics is a fairly new theory that has revolutionized the science of geology since the 1960s. It is really a theory about continually moving lithospheric plates that occasionally collide, scrape, and grind into each other as they move about the earth's surface. Prior to the modern theory, there were several other hypotheses, most notably *continental drift*, that hinted at a dynamic, changing earth.

According to research by marine geophysicist Roger Larson, changes in ocean plate activity may have a direct connection to sea level. Larson wrote that the seas which flooded the North American continent in the past may have been caused by a sudden increase in spreading at the ocean ridges.[2] An increase in the worldwide spreading rate would uplift the ocean floor due to increased heat flow, causing a rise in sea level and flooding of the land surfaces. Modern creationists have suggested that a similar process may have been one of the causes of the Flood.[3] They argue that a rapid increase in seafloor spreading rate, accompanied by the bursting of the "fountains of the great deep," caused the ocean crust to rise, pushing water upward onto the continents. Months later, when spreading slowed or ceased and the "fountains" were shut off, the ridges cooled and sank, allowing the ocean water to returned to the modern ocean basins and end the Flood.

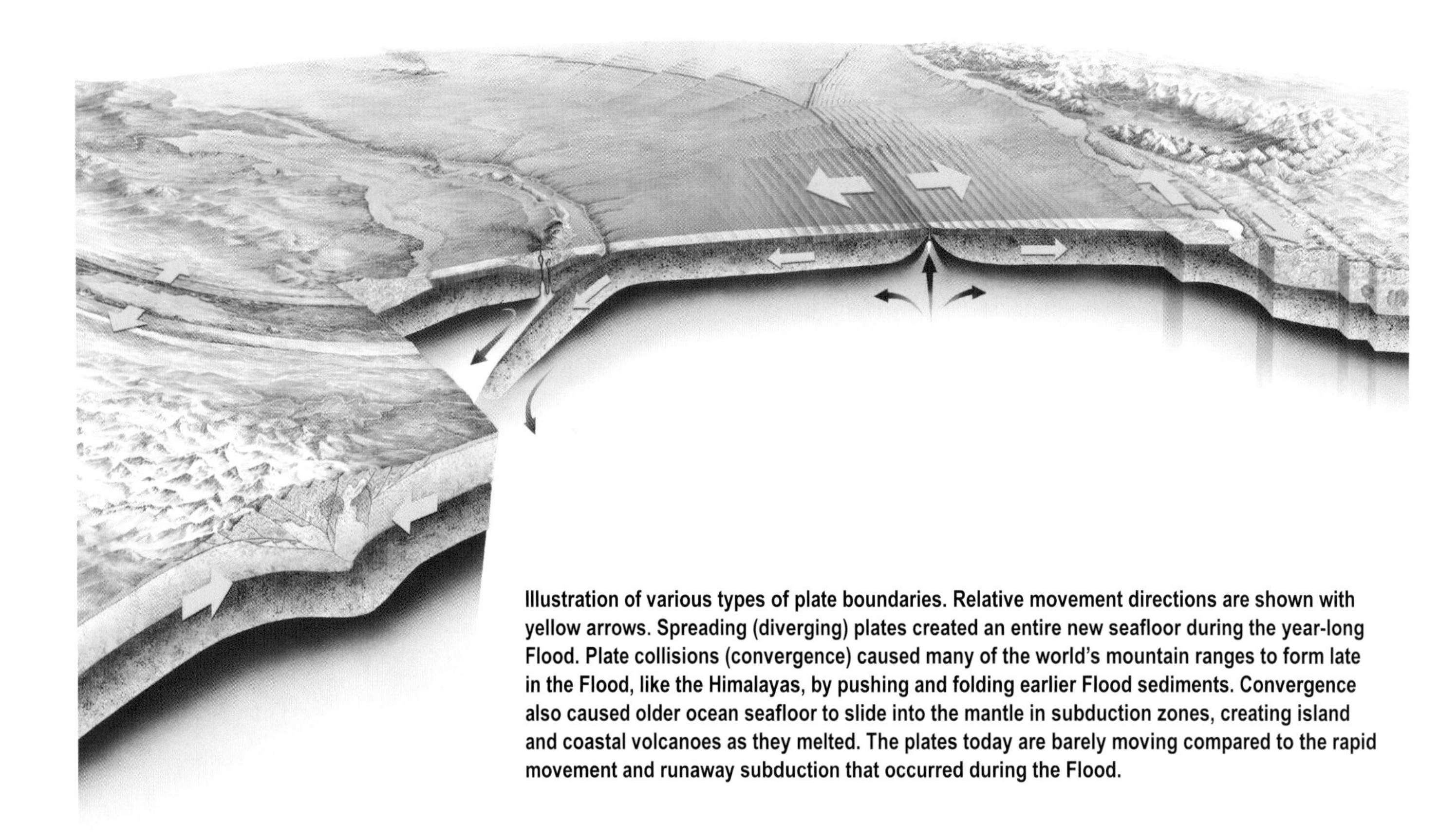

Illustration of various types of plate boundaries. Relative movement directions are shown with yellow arrows. Spreading (diverging) plates created an entire new seafloor during the year-long Flood. Plate collisions (convergence) caused many of the world's mountain ranges to form late in the Flood, like the Himalayas, by pushing and folding earlier Flood sediments. Convergence also caused older ocean seafloor to slide into the mantle in subduction zones, creating island and coastal volcanoes as they melted. The plates today are barely moving compared to the rapid movement and runaway subduction that occurred during the Flood.

Dinosaur fossils have been found on every continent, including Antarctica.

Creation geologists also believe earthquakes generated during the movement of the plates would have created huge tsunami-like waves to wash over the inundated continents. These waves would have spread the same sedimentary layers across great expanses of the continents. This process likely contributed to the different megasequences found stacked on the continents, as discussed later in this book.

EURASIA
NORTH AMERICA
PACIFIC
SOUTH AMERICA
AFRICA
PACIFIC
INDIA
ANTARTICA

One version of the reconstructed supercontinent of Pangaea.

Although there is a long list of scientists who proposed some sort of fit and subsequent breakup of the continents, one of the very first was Sir Francis Bacon. In 1610, just a little over 100 years after Columbus discovered the New World, Bacon noted the fit of the South American and African coastlines. However, the first to offer supporting evidence, other than the fit of coastlines, was German meteorologist Alfred Wegener. Between 1910 and 1930, Wegener proposed his theory of continental drift and backed it up with fossil and geological evidence. His data indicated the presence of a supercontinent he called Pangaea, which subsequently broke into the modern continental masses. Wegener believed the northern continental mass, which he called Laurasia, initially separated from the southern continents, which he called Gondwanaland. Not surprisingly, Wegener's hypothesis was not well received by mainstream geologists of the time and his research was largely ignored.

After Wegener, there weren't a lot of advocates in support of moving continents until 1962 when the hypothesis of *seafloor spreading* was introduced by Dr. Harry Hess. Because Hess was conscious of Wegener's failures, he introduced his paper as "geopoetry," so that he did not alienate his geologic audience. His evidence for seafloor spreading came primarily from his research related to the ocean ridges and studies of the ocean floor, driven by the Cold War efforts. Hess concluded that the ocean ridges were where new seafloor was being created from upwelling mantle material. This, he argued, caused the seafloor to spread in a conveyor-like fashion, pushing the ridges apart as new magma intruded from below, creating new ocean crust.

By the late 1960s, the modern theory of plate tectonics was born. It really developed as a combination of both continental drift and seafloor spreading, using the geologic data from the continents and the oceanographic data from the seafloor. Many scientists quickly collaborated on the formulation and publication of numerous papers, cementing the theory of plate tectonics in mainstream geology. The plates (or lithosphere) consist of the outer 62 miles (100 km) of the earth, and are composed of both crust and uppermost mantle. Most earthquakes, volcanoes, and mountain belts are associated with the boundaries between the various plates. Creation scientists have shown that rapid movement of the plates can account for the current separation of the continents within the year-long time frame of the Flood of Noah.[4]

Some Christians have suggested that the breakup of the pre-Flood supercontinent occurred after the Flood, referring to Genesis 10:25 (which refers to the earth being divided). Geologically speaking, this is unlikely. The tremendous tectonic forces and catastrophic effects necessary to break-up a supercontinent would have undoubtedly been devastating to all life on earth. More likely, the earth was "divided" in the time of Peleg by language differences and by God's scattering of mankind over the face of the earth (Gen. 11:9).

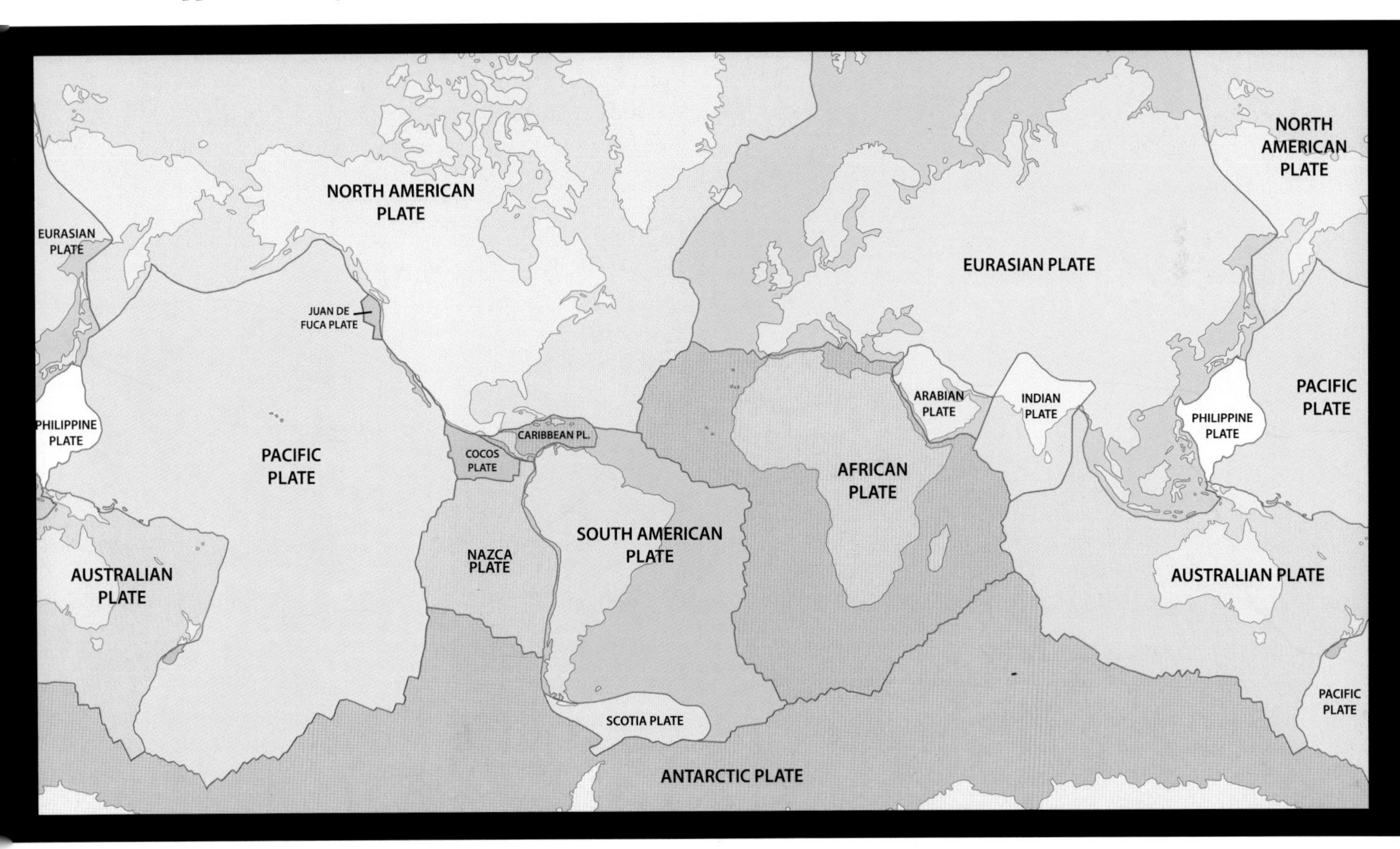

THE PRE-FLOOD WORLD

What was the pre-Flood world like? What do the fossils tell us? We do know that some types of dinosaurs grew to immense sizes, as shown by their fossil bones. But how did they attain these great sizes? Studies by uniformitarian scientists have suggested there was a very different set of environmental conditions in the geologic past, which creationist scientists apply to the pre-Flood world. Nearly all scientists agree there was once a supercontinent in the past, where most of life resided, including the dinosaurs. This one land mass would have allowed dinosaurs to attain the worldwide distribution we see today.

Computer models based on geochemical signatures found in rocks have shown conflicting results, but most indicate higher CO_2 levels persisted in the pre-Flood world (conventionally called the Paleozoic and Mesozoic Eras). Some have suggested CO_2 levels during this time were as much as three to seven times higher than today. A few paleontologists have also suggested that oxygen levels were much higher during this time, possibly making up as much as 30 percent of the (pre-Flood) atmosphere. Numerous studies have shown that gigantism in animals is connected to higher oxygen levels. Paleontologists simply point to the gigantism exhibited by fossil insects (dragonflies with wingspans of two feet) and dinosaurs (up to 160 feet long) to support these conclusions.

Higher concentrations of oxygen and carbon dioxide in the pre-Flood world would have allowed plants to flourish and helped animals grow to tremendous sizes. In addition, there was most likely a worldwide greenhouse effect due to the higher concentrations of CO_2, providing a warm, and likely moist, climate everywhere. We see fossil evidence of lush vegetation and a diverse distribution of animals across all continents, including coal beds and dinosaurs in Alaska and Antarctica.

As will be discussed in chapter 12, the extinction of dinosaurs may have been partially caused by rapid climate changes and shifts in atmospheric composition after the Flood. A significant drop in atmospheric oxygen concentration to the current 21 percent, combined with cooler conditions in the north and south latitudes, would have limited the range of the dinosaurs and possibly slowed their activity levels, especially if they were cold-blooded.[5]

The Bible indicates there was little violent weather before the Flood, and that the surface of the earth was watered primarily from below through a different system of hydrologic conditions (Gen. 2:6). The scarcity of severe storms and associated lightning may have been necessary to keep forest fires from raging if oxygen levels were as high as 30 percent. God's providence abounded everywhere.

At the end of the Flood, God made a covenant that He would never again use a worldwide flood to destroy every air-breathing creature, including the dinosaurs (Gen. 9: 13–17). The sign for that covenant is the rainbow. Some Christians have argued that the Flood of Noah was only a local event, and not global. These verses in Genesis chapter 9 refute this claim. If the rainbow was a promise to never again send a "local" flood upon the earth, God would have broken this promise countless times with every swollen river or tsunami event.

> I have set my rainbow in the clouds, and it will be the sign of the covenant between me and the earth. Whenever I bring clouds over the earth and the rainbow appears in the clouds, I will remember my covenant between me and you and all living creatures of every kind. Never again will the waters become a flood to destroy all life. Whenever the rainbow appears in the clouds, I will see it and remember the everlasting covenant between God and all living creatures of every kind on the earth." So God said to Noah, "This is the sign of the covenant I have established between me and all life on the earth."
>
> —Genesis 9: 13–17 (NIV)

PLANTS CREATED ON DAY 3

What were the plants like in the pre-Flood world? Are the paintings and drawings in books and the movies accurate? Again, if we turn to the rocks and fossils, we may find some answers to these questions.

Paleontologists have found evidence that dinosaurs ate nonflowering plants like ferns and conifers, finding them in the stomach region of well-preserved fossils of hadrosaurs (duck-bill dinosaurs). Scientists have also found that sauropods (long-necks) ate grass like the behemoth in the Book of Job (Job 40:15), finding fossilized grass (angiosperm) remains in their dung.[6] Even a few angiosperm (flowering plants) fossils have been found in Flood sediments as deep as the Middle Triassic system, coinciding with the deepest discovered dinosaur bones. So it appears that at least some flowering plants were growing in the areas of the pre-Flood world where the dinosaurs resided.

However, the vast majority of fossil plants found with dinosaur remains are nonflowering plants, implying that the dominant plant types were quite different in the pre-Flood, "dinosaurian" world. The warmer climate and possible atmospheric differences allowed the pteridophytes (seedless vascular plants, like ferns) and gymnosperms (naked seeded vascular plants, like conifers) to dominate the landscape in the areas where the dinosaurs lived. As we shall discuss later in chapter 14, the dinosaurs may have lived in lowland areas closer to sea level, which may have provided a unique ecosystem for nonflowering plants to flourish. For sure, there were some flowering plants (angiosperms) in this environment, but they were much less dominant, and probably mostly grasses that didn't fossilize easily.

Most evolutionary paleontologists believe flowering plants, or angiosperms, did not "evolve" until the Lower Cretaceous, supposedly 135 million years ago. Dinosaurs are usually only pictured among ferns, conifers, and cycads plants when evolutionary scientists reconstruct their pre-Cretaceous world. They often refer to the Cretaceous as a time of transition when the world climate supposedly changed, and the plants and even the dinosaurs changed as well. Still, the rather sudden appearance of flowering plants in Cretaceous system strata has been difficult for evolutionary scientists to explain right from the start. Charles Darwin referenced the sudden appearance of fully formed flowering plant parts in the fossil record as an "abominable mystery" in a letter to Joseph Hooker in 1879. He found no ancestors to these plants. Creation scientists have always insisted that all plants, including the flowering plants, were created together on day 3 of creation week just as the Bible reports (Gen. 1:11).

Scientists were surprised recently when they discovered angiosperm-like pollen-grain features from Middle Triassic system rocks in Switzerland, pushing the first appearance of flowering plant fossils much deeper than conventionally thought.[7] Tests indicate it came from a

flowering plant rather than a seed-producing plant like conifers and cycads.

Many earlier reports, going back to the 1960s, of angiosperm fossils found in pre-Cretaceous strata have been ignored or explained as some sort of contamination. Unfortunately, most of these findings received mixed interest by the evolutionary community, possibly because pre-Cretaceous angiosperms disrupt their well-established evolutionary tale.

For example, Dr. Robert Bakker devoted an entire chapter in his book *The Dinosaur Heresies* to "When Dinosaurs Invented Flowers." Bakker explained, "When plant-eating dinosaurs evolved more effective teeth or fermenting chambers, the plant species had to adjust to the new weaponry or die." Bakker continues, "As more and more new kinds of Cretaceous beaked dinosaurs entered the system, more and more angiosperm families evolved."[8]

According to evolutionists, grass was not even supposed to be found in Cretaceous strata. It was claimed to have evolved after the dinosaurs went extinct. However, in 2005, grass fossils were discovered in sauropod dinosaur coprolites (dung) from India.[9] Dinosaurs were never pictured with grass in their mouths, but now they certainly should be. The discovery of angiosperms in rock layers below those containing most dinosaurs, and the discovery of grass fossils among sauropod (long-necked) dinosaurs confirm the biblical details about behemoth, the dinosaur from Job 40 whose tail "sways like a cedar." The text describes its diet, saying, "Look at Behemoth, which I made along with you and which feeds on grass like an ox." Prior to the grass particles in fossil dinosaur dung from India, the evolutionary picture had no place for dinosaurs eating grass, but it was in Scripture all along. The text affirms that God made behemoth "along with you," meaning during the creation week when God made angiosperms on day 3 and dinosaurs and then man on day 6. So it's no surprise for biblical creationists to find flowering plant fossils mingled with animal fossils. The Bible tells us they lived together, and science confirms this fact.

The Bible's record of the major phases of world history shows no trace of Triassic, Jurassic, or Cretaceous time intervals. Scripture reports the instant creation of each plant kind. Furthermore, the Bible describes in detail a worldwide Flood capable of preserving life's traces in fossil forms. In that context, Triassic, Jurassic, and Cretaceous flora and fauna do not represent a separate time but distinct ecosystems buried by sediment-laden floodwaters. The Bible's timeline shows a creation that is thousands, not billions, of years old, erasing any need to explain why pollen grains buried deep in fossil layers look so similar to living herbs and flowers today.

TYPES OF DINOSAUR FOSSILS

Scientists have found many types of dinosaur fossils from bones, skin imprints, eggs, footprints, and gastroliths (gizzard stones) to coprolites (feces). Fossils are defined as the remains and/or traces of ancient life, generally assumed to be Ice Age or older. The study of fossils is called paleontology. It is similar in some respects to archaeology. However, archaeology concentrates on human artifacts, like pottery and Egyptian mummies, and paleontology studies the fossils buried in sedimentary rocks.

Few "fossils" are forming today and are rarely, except in extreme conditions, forming anywhere on earth. And yet there are billions of fossils scattered all around the world, from Antarctica to the Asia. They are found in many types of sedimentary rock and all through the many layers, but are particularly noticeable within the Cambrian strata and above. That is why evolutionists have designated this layer as the "Cambrian Explosion." It marks the deepest layers where we find a plethora of fossils. They still struggle to explain why most major animal phyla just suddenly appear at this level, worldwide, fully formed and functional, without any trace of ancestry. This remains a major unsolved mystery in secular science.

In contrast, creation scientists have provided a more satisfying explanation for the Cambrian Explosion event and all subsequent fossil-rich layers. The Cambrian layer is simply interpreted as the beginning of the Flood deposition that spread across each of the continents. This layer buried all sorts of animals and plants rapidly, many while still alive, preserving them for us to discover. It is also part of the first recognized megasequence, termed the Sauk, discussed in more detail later in this book. I will also explain in chapter 12 why dinosaurs are not found in these earliest Flood layers.

As an example of the number of fossils in just one layer, let's look at some research from Montana. Jack (John R.) Horner and James Gorman, in their book *Digging Dinosaurs*, described a single hill in Montana which contains the remains of approximately 10,000 individual *Maiasaura* and over 30 million fossil fragments.[10] *Maiasaura* is a plant-eating, duck-billed dinosaur that is found buried in Cretaceous system rocks (deposited later in the Flood).

Remember, special conditions are required to preserve and make a fossil. Things don't just die and become a fossil with the right mix of "time." Today, scavengers first eat the choice parts of any dead animal, then aerobic and anaerobic bacteria complete the decomposition process. "Dust to dust" is an appropriate phrase in most cases.

How, then, are things fossilized? The most important condition is a rapid burial. Catastrophic activity is a prerequisite. Where and when does this occur? Large floods are most commonly envisioned when it comes to land animals and plants. However, landslides, sometimes set off by earthquakes or heavy rains, can also bury critters sufficiently. Just sinking to the bottom of a lake or sea is not enough, as there are scavengers and aerobic bacteria there too. So we not only need a rapid burial, we need a rather deep burial to cut off the oxygen supply or at least limit the oxygen supply. This step is necessary to slow and prevent complete disintegration by bacteria. Most evolutionary scientists think dinosaurs were catastrophically buried during rainy seasons and/or river flood events. They believe the stacked layers of fossils are the result of repetitive flood events separated by an indiscriminate amount of time between events. They ignore the lack of evidence in the rocks to support their "missing" time between the layers, and disregard the truth in the Book of Genesis, which explains how billions of fossils formed in one, year-long Flood that buried them all, in rapid succession.

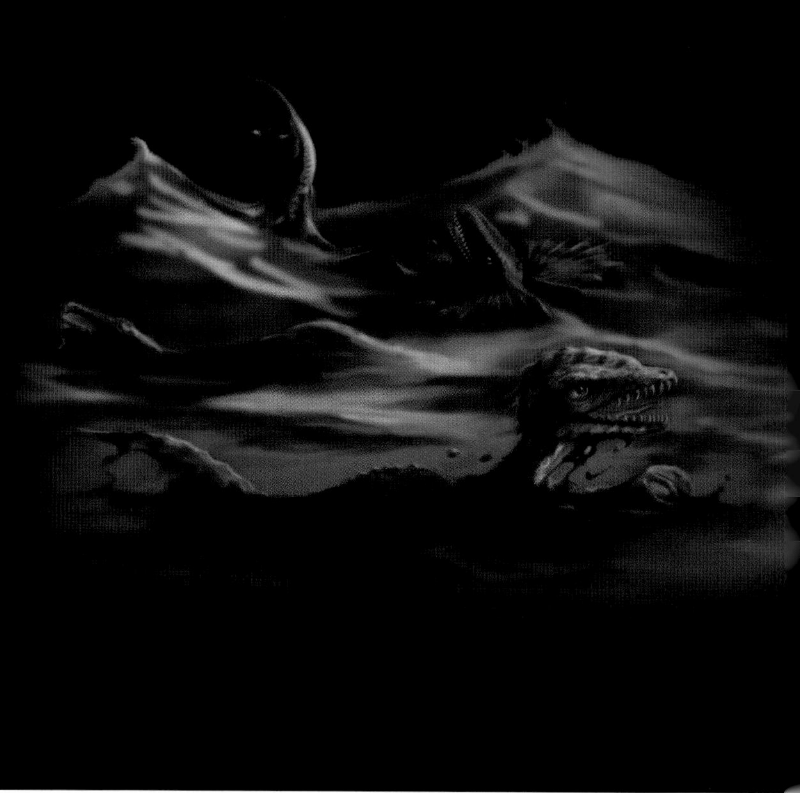

Fossils are found in many types of sedimentary rocks, but are most common in shales (mudstones) and limestones. They are less common in sandstones, even though most dinosaur fossils are found in sandy layers associated with high energy sand movement. Shales and limestones allow better preservation of small fossil and minute details as the rocks are usually finer grained. Second, rapid groundwater flow usually dissolves most critters buried in sand prior to lithification. Sand is very porous, and more groundwater is able to flow through sand, compared to clay and lime mud, lessening the chance of preservation.

Most secular geologists feel that the fossil record (the collective information gained from all discovered fossils) is incomplete. Part of the reason they believe this is the selective nature of fossilization. Hard parts are obviously easier to preserve than soft parts. Therefore, the soft tissue and inner organs of most vertebrates are rarely found. This also carries over to marine invertebrates, where only the outer shells are typically preserved as fossils. The fossil record is obviously biased in favor of hard shells and bones. In fact, many mammals are known and identified only from their teeth because enamel is so strong and durable compared to the other bones.

Finally, many people get the wrong impression that fossilization takes millions of years to occur. Evolutionary scientists need lots of time to try and explain evolutionary changes that they believe happened. Pyritization (the petrification by replacement with the mineral pyrite) of plants can be an extremely rapid process, taking as little as 80 days, under the right conditions.[11] Remember, just because it's a fossil doesn't make it millions of years old. As we will see later, the findings of preserved soft tissues in more and more fossils indicates the fossils are only thousands of years old.

There are many types of dinosaur fossils. Most can be lumped into the following general varieties: 1) petrification (which includes permineralization and replacement), 2) preservation, 3) carbonization, 4) molds and casts, and 5) trace fossils.

Petrifaction means to change the original organism to stone. This can be accomplished in several ways, but usually involves the removal of the majority of the organic material associated with the dead organism. Groundwater typically dissolves away the original parts while simultaneously depositing inorganic minerals, preserving many of the details of the dead organism. These deposits, such as quartz and pyrite, leave behind exact replicas of the fossil for geologists to study. Many dinosaur bones have experienced partial or full petrification, sometimes even being replaced with radioactive uranium.

Preservation is the retention of original organic parts. This type of fossil is less common, but scientists are finding more and more preserved dinosaur soft tissue all the time. Even the enamel in dinosaur teeth may be preserved. Preserved soft tissue has been found in scores of fossil specimens claimed to be many millions

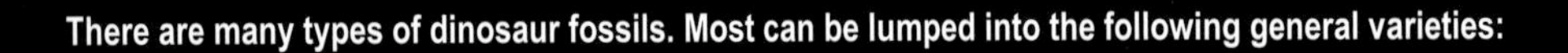

There are many types of dinosaur fossils. Most can be lumped into the following general varieties:

petrifaction
(which includes permineralization and replacement)

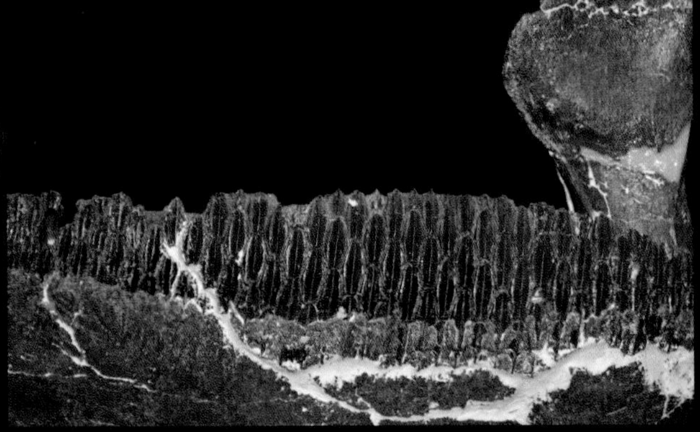

preservation (teeth enamel is commonly preserved)

of years old, including numerous dinosaurs, since 2005, from embryos of a sauropod to the leg bone of a *T. rex*.[12]

Carbonization is a typical type of fossilization for plants. It is less common in dinosaurs, but there have been some reports of dinosaur internal organs preserved as a carbon residue. Commonly, plant leaves and other parts of a plant are buried, and during lithification the organic material "cooks" or distills and leaves behind a thin, black carbon residue.

Molds and casts are the impressions or "shapes" of ancient life. Often the actual organism dissolves away, leaving only the imprint of the organism in the rock. Molds are created in this way. There can be internal molds, which give the internal imprint of an organism, and external molds, which give the imprint of the outside of the organism. Dinosaur skin prints are a type of external mold. Casts are a secondary process, where a mold must first be created by dissolution of the original shell or bone and the cavity is filled in with sediment or other minerals.

Trace fossils include tracks, burrows, trails, and anything else that indicates activity of an organism, without being a direct part of the organism. Other trace fossils include coprolites, or fossilized feces, and gastroliths, or gizzard stones, that helped some dinosaurs digest their food.

Fossil of *Acanthoteuthis*, an extinct squid, at the Museo di Arsago Seprio.

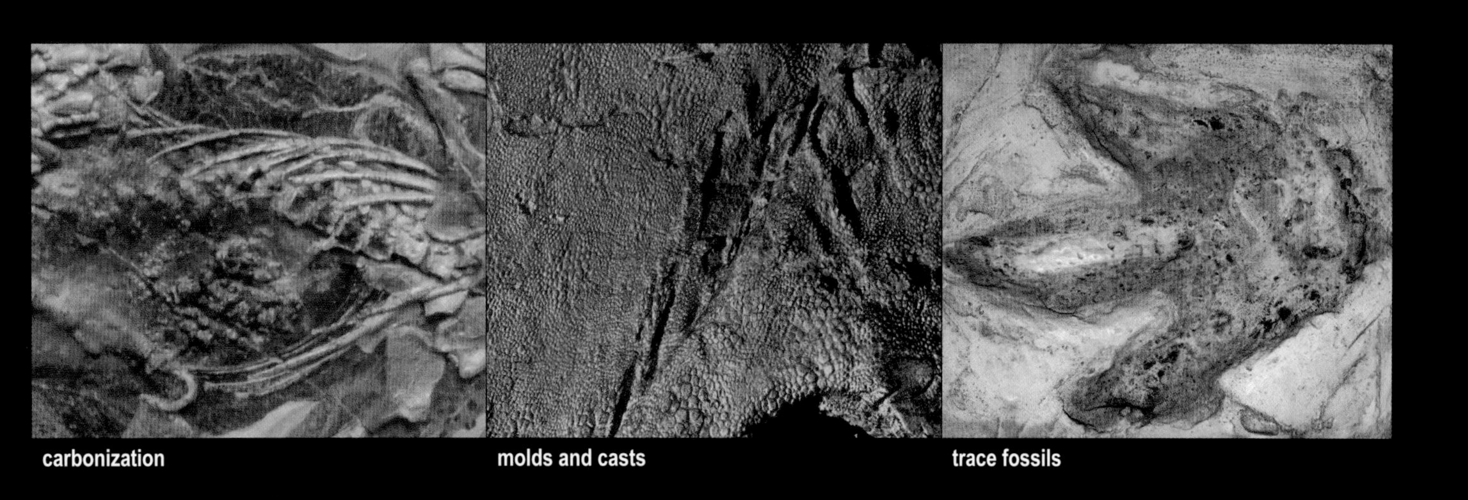

carbonization | **molds and casts** | **trace fossils**

Dinosaur fossil found buried along with a fish

IN WHAT TYPES OF ROCKS ARE DINOSAURS FOUND?

Secular geologists describe two major environments where sedimentary rocks are thought to form:

1. marine or ocean environments under the sea
2. continental or terrestrial (nonmarine) environments on land.

Within each of these two broad categories exists multiple smaller varieties of deposition, such as shallow or deeper water locations. When looking for dinosaurs, secular geologists look almost exclusively in the continental sedimentary rocks, as dinosaurs didn't live in the oceans. However, during the Flood there were large expanses of ocean sediment and marine organisms pushed rapidly onto the land, placing what was a marine environment onto the land. Because of this catastrophic activity, dinosaurs are nearly always found mixed in with marine fossils. The Hell Creek Formation in Montana contains the remains of numerous tyrannosaurs and also 5 species of sharks, 15 species of fishes, 17 species of mammals, 17 species of turtles, and various lizards and amphibians.[13] Secular scientists are left claiming all these sharks and fishes are freshwater varieties. They refuse to admit the possibility of Noah's Flood creating this odd mix of fossils.

Most dinosaur fossils are found in massive bones beds where thousands of bones of different dinosaurs are found together, usually torn apart or disarticulated, in a sort of log jam. Rarely are dinosaurs found as complete animals. Bone beds are factual evidence supporting a catastrophic burial for the dinosaurs. Dr. Art Chadwick has reported between 10,000 and 25,000 individual dinosaurs in a single bone bed in the Powder River Basin of eastern Wyoming. He and his students excavate thousands of bones every season from this site. He envisions the deposit as a Flood-related debris flow that poured the dinosaurs into a shallow layer on the continent. Today, the bones are found mixed with marine fossils in a 20 inch thick (0.5 m) mudstone layer in the Lance Formation.

Paleontological Museum in Berlin

CHAPTER 3:

THE TRUE AGE

Dinosaurs are found in rock layers that secular geologists have labeled as the Triassic, Jurassic, and Cretaceous systems, from lowest to highest, respectively. In many cases, different dinosaurs and other creatures are found in each different system. Why is this? The various segments of the geologic column, including these rock layers, were divided and named based on the distinct fossils found within them. In other words, Jurassic rocks layers tend to have different fossilized organisms than are found in Cretaceous rock layers. Each of these major rock layers (i.e., Jurassic) is called a "system."

Black basalt lava from the Kilauea Volcano, Hawaii Island (Big Island), Hawaii.

Systems are further divided into series and then stages, based on increasingly subtle differences in the index fossils. A system itself is a subdivision of a larger unit called an "erathem." All dinosaurs are found within the same erathem: the Mesozoic, which is divided into the Triassic, Jurassic, and Cretaceous systems. A possible Flood-related explanation of these subdivisons, and apparent extinctions, is discussed later in this book.

Secular geologists believe that each system was deposited over millions of years. Unfortunately, they label the system a time period, using the same name as the rock layers that they think were deposited at that time (believing that Triassic system rocks were deposited during the Triassic period). This terminology can lead to some confusion. Only context will make this clear. In secular thinking, "eras" are large divisions of time that correspond to erathem rock layers, "periods" subdivide eras and correspond to systems of rocks, "epochs" subdivide periods and correspond to the series level, and "ages" subdivide epochs and correspond to stages of the geologic column. Secular scientists believe all these subdivisions represent "eras, periods, and epochs" of time that are bounded top and bottom by extinction events.

As we examine the geologic column, moving from lower to higher systems (earlier deposition to later deposition), we find that every fossil appears fully formed, showing little change with position in the column (demonstrating stasis), and then disappears just as suddenly. Some fossils appear only in one system, whereas others are found in many systems. There are few, if any, undisputed transitional fossils to link species. Secular paleontologists believe these changes in the fossils are the result of evolution and extinctions that took place over short segments of geologic time.

The K-T (Cretaceous-Tertiary) or K-Pg (Cretaceous-Paleogene, as it is now called) extinction event is an example of one of these. Since nearly all dinosaur fossils are found below the Cretaceous-Tertiary boundary level, secularists believe that dinosaurs went extinct at the end of the Cretaceous *period*. Creation scientists interpret these changes from one system to another not as examples of extinctions, but as differences in tectonic activity, depositional energy, and changes in current direction during the one great Flood event.

The Bible tells us that the Flood began with the bursting of the "fountains of the great deep" (Gen. 7:11). These large cracks, or rifts, in the earth appeared all over the planet, simultaneously. The rocks left behind from this activity were mostly vast buildups of black, basaltic lava, similar to what is observed in Iceland or Hawaii. One such rift runs down the center of the continental United States from Lake Superior to central Kansas, and is called the Midcontinent Rift. This may have been one of the "fountains" that began erupting as the Flood initiated. Other rifts appeared around the globe. Some of these progressed into new oceans, and their remnants are still visible today as rifts along the center of the modern ocean ridges.

> In the six hundredth year of Noah's life, on the seventeenth day of the second month—on that day all the springs of the great deep burst forth, and the floodgates of the heavens were opened. And rain fell on the earth forty days and forty nights.
>
> —Genesis 7:11-12 (NIV)

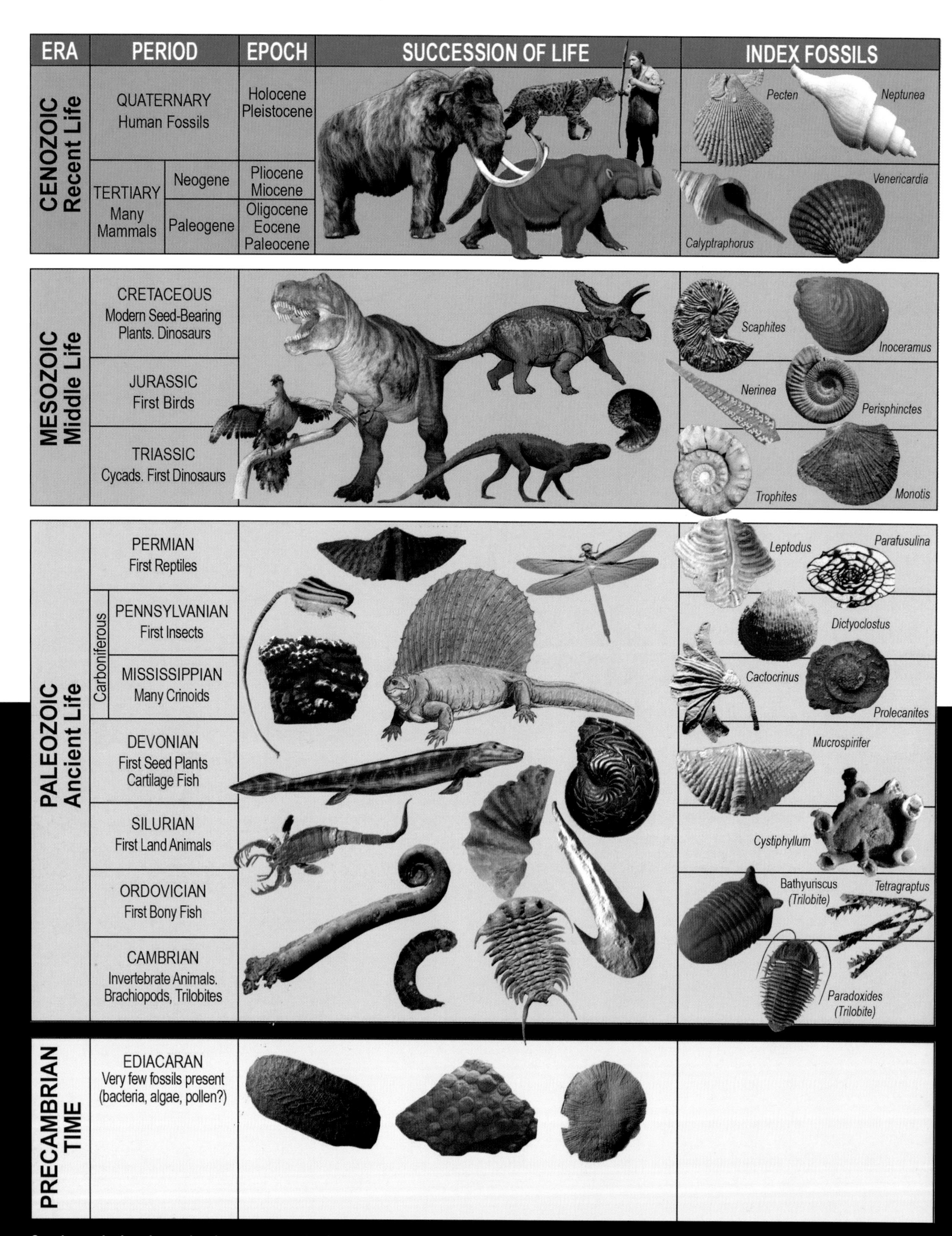

Secular geologic column showing representative fossils found in various strata.

Along with the lava, the upper mantle or lower crust apparently produced tremendous quantities of superheated groundwater that also poured out of the many rifts. As water and lava burst forth, some of the nearby crust seems to have collapsed into the void space left behind.

Some of the earliest Flood deposits (rocks composing the Paleozoic erathem) contain mostly shallow marine fossil organisms. Study of hundreds of stratigraphic columns across the continental United States and Canada shows many of these early Flood deposits thickening dramatically in the Michigan, Illinois, Permian, Hudson Bay, and Williston Basins. These so-called intracratonic basins are likely a consequence of withdrawal of magma and water from nearby rifts and the resulting crustal subsidence. The initiation of catastrophic plate tectonics and subduction early in the Flood probably caused tsunami-like waves to sweep across the lower levels of the continent, burying millions of marine animals in the deposits. This interpretation may explain why the earliest Flood layers contain mostly shallow marine organisms with few large animals and few land animals. The early Flood water levels may not have reached the land animals, or the land plants, until later.

As the Flood progressed (depositing what's called the Mesozoic erathem), the stratigraphic columns indicate a major shift occurred in depositional style. At this point, it appears earth's tectonic plates began to move even more violently, and the process of subduction began to consume the pre-Flood ocean crust along the U.S. West Coast. The U.S. East Coast ceased subduction and began to break and rift away from Africa and create the Atlantic Ocean at several feet per second. This change in tectonic style released even greater and more massive waves to wash the bigger and deeper marine animals onto the continents, including large marine reptiles like plesiosaurs and mosasaurs. It was the higher energy waves that also swamped the dinosaurs in rapid succession and mixed them in with marine organisms. We also see evidence of this change in tectonic style in the rocks themselves, with large sand waves indicated by huge cross-bedded sandstones in the Permian and Jurassic systems in North America.

Imagine water and sediments and organisms from each wave cycle suddenly piling on top of sediment and organisms from previous cycles as huge tsunami-like currents shifted back and forth during the Flood. This would result in a "sorting" of the fossils in the rock layers just like we see in the geologic column. As the Flood progressed, animal and plant communities from different depths, levels, and environments would become placed on top of one another, resulting in completely different fossils suddenly appearing and disappearing, like the rocks show today.

The succession of fossils in the rocks is not a succession of evolution through time. Rather, it is a record of the Flood history, shifts in tectonic style, and changes in energy and current directions, giving us a chapter-by-chapter "book" from which to read the record of God's watery judgment!

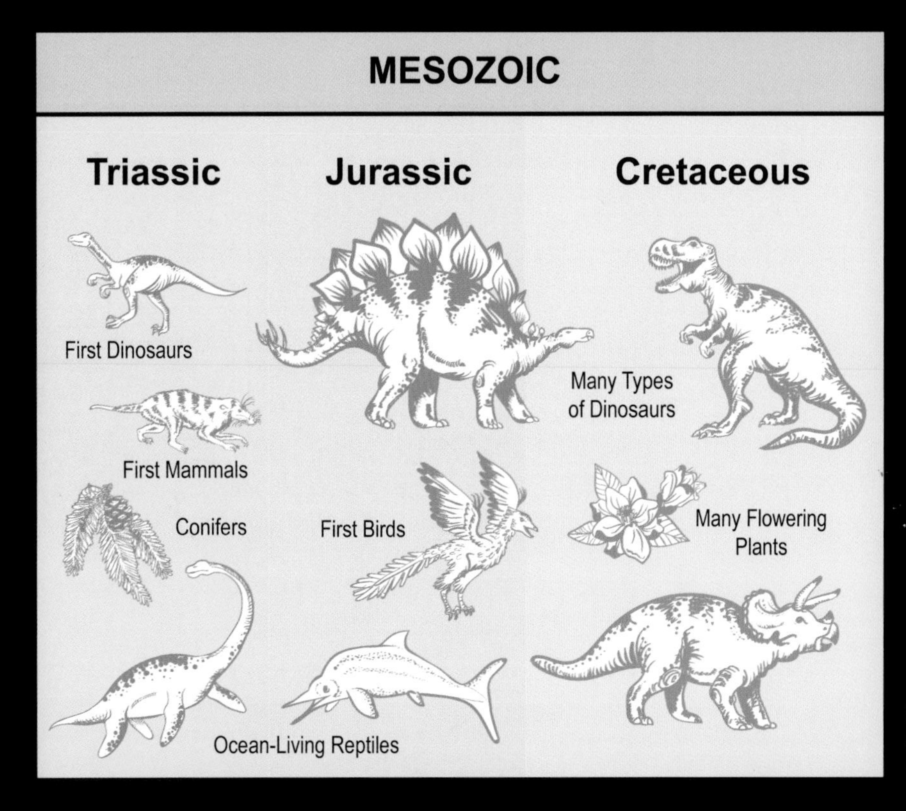

Secular chart showing the three systems of the Mesozoic and representative fossils.

It is noteworthy to realize that geologic systems are real. We can observe that certain fossils are found higher or lower than other fossils, so there is no doubt about the existence of the Triassic, Jurassic, and Cretaceous *systems*. However, this does not in any way prove the existence of Triassic, Jurassic, and Cretaceous *periods*. We cannot tell from observations alone when these rocks were deposited. The stratigraphic column is real. The geologic time span is not.

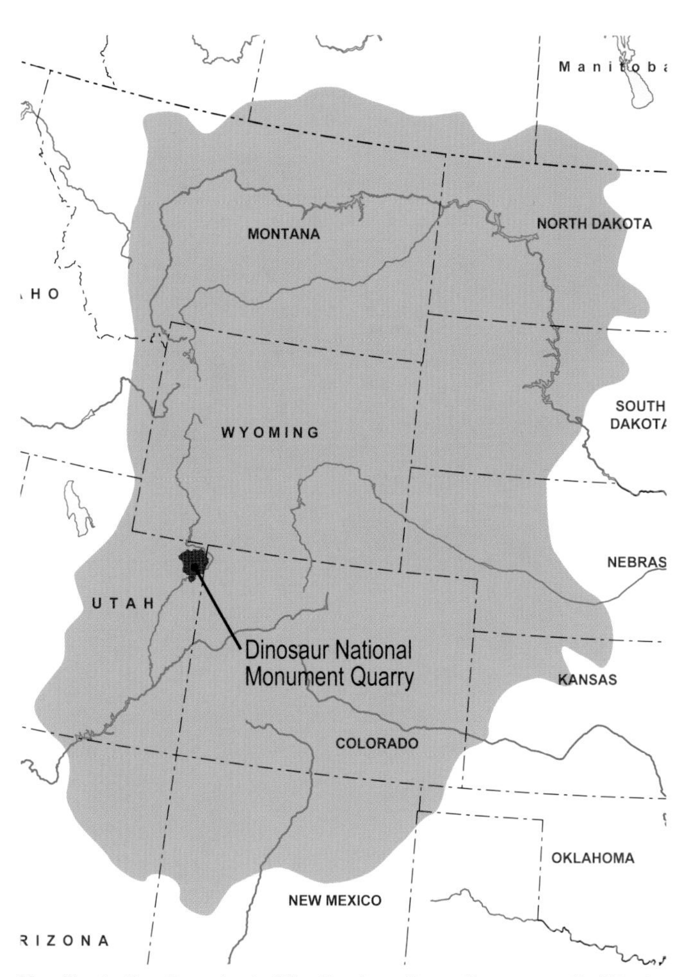

Map illustrating the extent of the Morrison Formation across the West.

Secular geologists always try to interpret the layers of sedimentary rocks into one of today's known environments by what patterns they observe in the rocks. This becomes an exercise in uniformitarianism, believing that each ancient rock layer can be explained by a modern environment that is observed in the present. They commonly show dinosaurs being deposited in ancient river systems or river deltas because they see the rock evidence for water deposition and "floods" to bury dinosaurs rapidly. However, these interpreted rivers and deltas are not exactly the same as today's systems. Many ancient "river channels" resemble today's deep ocean sand channels created by ocean currents, some of which even meander like rivers, but occur deep beneath the surface.

The Jurassic Morrison Formation is interpreted by secular scientists as the result of ancient rivers that slowly migrated across much of the American West over millions of years. But the idea of simultaneous rivers, spreading an equal amount of sediment in a single layer across thousands of square miles, encompassing numerous states, makes little logical sense. Creation scientists interpret this

Outcrops of the scenic multicolored Morrison Formation.

Stegosaurus (left) and *Tyrannosaurus* (right).

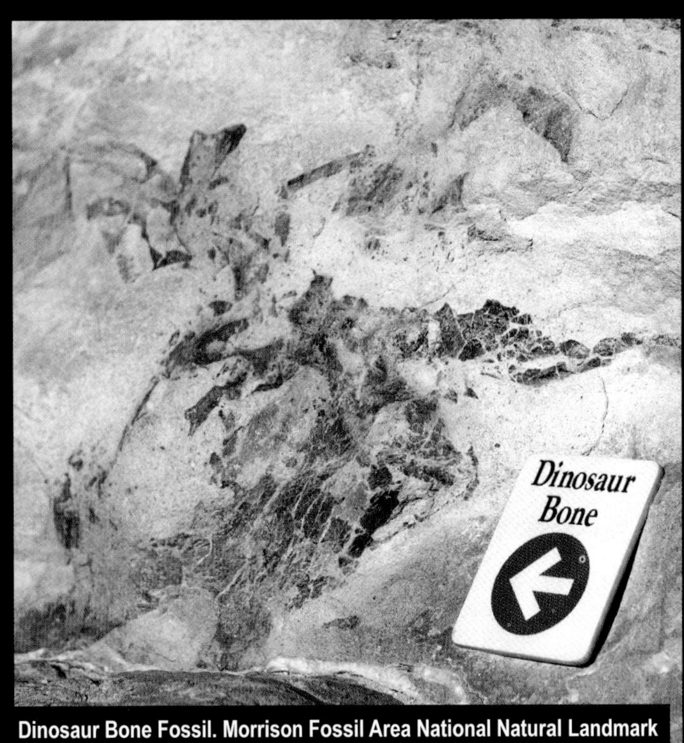

Dinosaur Bone Fossil. Morrison Fossil Area National Natural Landmark

as an extensive and rapidly deposited unit of sands and clays that occurred during the great Flood thousands of years ago, trapping dinosaurs and other organisms while many were still alive. Thousands of dinosaur bones of many species were excavated from the Morrison Formation bone beds near Como Bluff and Thermopolis, Wyoming, and from numerous areas in Utah and Colorado. Some of the species found within this unit include *Apatosaurus*, *Allosaurus*, *Stegosaurus*, and *Camarasaurus*. The Morrison Formation continues to be the source of discovery for many dinosaur fossils. It is named after the small town of Morrison, Colorado, where the rock unit was first described, and dinosaurs were also found nearby. The great extent of the deposit is strong evidence of the expanse of the worldwide Flood.

Why have *T. rex* and *Stegosaurus* never been found together in the same rock unit? *T. rex* is found only in what is called Upper Cretaceous rocks, and *Stegosaurus* is found only in Upper Jurassic rocks (Morrison Formation). It's likely that these two dinosaurs lived in completely different pre-Flood environments, and possibly at different elevations, so that the Flood did not deposit their remains with each other, but only on top of one another.

MAMMALS AND DINOSAURS?

It is common knowledge among paleontologists that mammals and dinosaurs are found in the same rock layers. Mammal bones and teeth are found in rock strata as deep in the Flood sediments as the Triassic system, coinciding with the first appearance of dinosaur fossils. Recall that creation scientists consider these strata all part of the sediments that were deposited during the great, year-long Flood. The depth of burial really is a consequence of Flood sediments piled on top of one another in rapid succession.

What is generally not known is the incredible diversity of mammals that have been found mixed in with the dinosaur fossils. We are usually told by evolutionary scientists that all the mammals found with the dinosaurs were small, shrew-like animals. As the story goes, these small mammals scurried around avoiding dinosaurs and hiding away in burrows until the dinosaurs presumably were nowhere in sight. Some evolutionists point out that reptiles, like the dinosaurs, would do better than mammals in the hot, dry climates they envision for the Triassic system, because reptiles have less porous skin and water-conserving kidneys. Other evolutionists suggest that these small "early" mammals were egg-layers like the modern platypus.

However, this tale is simply untrue and unsupported by recent discoveries. Mammals found in the same Flood sediments with dinosaurs show a great variety in style and size. Some of these mammals resemble modern squirrels and beavers, and others were over three feet long. There is even evidence that some of these mammals were eating the dinosaurs and not just the other way around.

Paleontologists have described a fossil mammal that closely resembles a modern beaver from Middle Jurassic system strata in China, claimed to be 164 million years old. The body length was estimated at two feet, and the fossil retained impressions of the guard hair and underfurs. This semiaquatic mammal also had forelimbs well-designed for digging and rowing. Called *Castorocauda lutrasimilis,* this species was found mixed in sedimentary layer with pterosaurs (flying reptiles), abundant insects, amphibians, a dinosaur, and a conch-like gastropod mollusk. This seemingly odd conglomeration of marine and nonmarine animals is best explained by a tsunami-like Flood deposit transporting ocean water onto the land and mixing different environments into one unit.

Many mammal fossils found buried with dinosaurs were small. One shrew-sized mammal, found in Transylvania, Romania, even had red teeth. *Barbatodon transylvanicus* was found in Upper Cretaceous strata about 120 miles from the ancient castle of Vlad Dracula. Thierry Smith and Vlad Codrea concluded that the red teeth came from enamel that was composed of 3 percent iron. It didn't suck blood from prey, however, but likely used its red teeth on insects, similar to modern shrews.

Duck-billed platypus in Rainforest Creek

***Castorocauda lutrasimilis*, a docodont from the Middle Jurassic of China**

Barbatodon transylvanicus

One of the largest mammals found in fossil beds with dinosaurs comes from Upper Cretaceous system strata near Liaoning, China, dated by evolutionists as 130 million years old. Two specimens of a Tasmanian devil-like animal (*Repenomamus*) were found at one site. The larger one was over three feet long. The smaller one (about 20 inches long) had the fossilized bones of a juvenile *Psittacosaurus* dinosaur preserved in its stomach contents. This was the first documented fossil of a mammal eating a dinosaur. The large size of these mammals and their predation of dinosaurs came as a complete surprise to evolutionary paleontologists and the stories they usually teach. Creation scientists were not shocked, however, to find that pre-Flood mammals were not all small, docile, shrew-like animals.

An *R. robustus* feeding on a *Psittacosaurus* hatchling.

We still have many of these same kinds of larger, carnivorous mammals living today.

Apparently, some dinosaurs did try to dig up burrowed mammals and hunted them for prey. Dinosaur claw marks and scratched surfaces, matching the feet claws of dromaeosaur theropods like *Velociraptor,* were found within fossilized mammal burrows in Utah. The paleontologists who studied the site concluded that the scratch marks were preserved as the burrows were rapidly buried by sand during a flood. Creation scientists know this is further confirmation of the great Flood and not just a random local flood event.

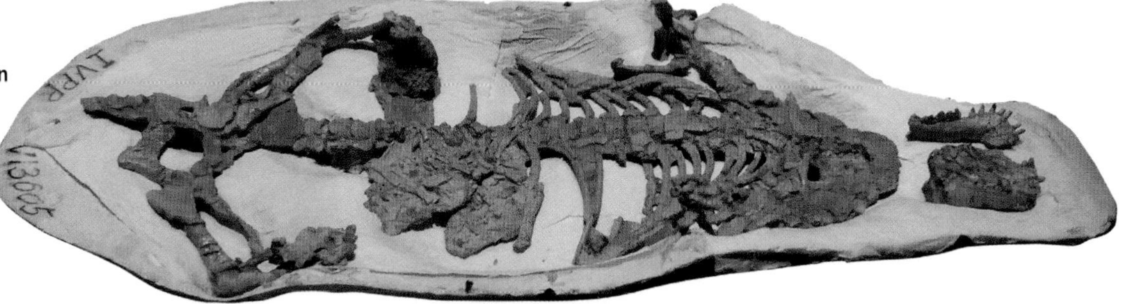

The *R. robustus* specimen with *Psittacosaurus* remains in its stomach

Giant Pleistocene Beaver, *Castoroides ohioensis*

Taken at the Minnesota Science Museum: Mississippi River Gallery.

Geologic time is estimated in two different ways. It is either viewed on a relative scale or on an absolute scale. Relative time places geologic events in an order from oldest to youngest, or vice versa. This type of time is only concerned with which events have occurred prior to other events. Creation scientists are attempting to work out the geologic history of the Flood using this method. Absolute time, by contrast, attempts to assign the number of years before present to each event. It is only possible to determine absolute time through historical records. However, secular scientists have attempted to determine the ages of rocks by analysis of radioactive isotopes.

In the late 18th century and into the 19th century, geologists began to develop a philosophy of uniformitarianism, where the earth's processes were viewed as unchanging and progressing at very slow rates. Inches of sediment were equated to thousands of years. Within the same century (1859), Charles Darwin published his now-famous treatise on evolution. Since natural selection does not change one kind of organism into a different kind on an observable timescale, Darwin had to add vast amounts of time for evolution to be viable. These scientists needed to make the earth older than the 6,000 years described in the biblical genealogies.

The earth quickly "aged" to about 20 million years old in the minds of secular scientists by the late 19th century. With the discovery of radioactivity in 1896 by Henri Becquerel, the earth was allowed to "grow" even older, and estimates of age greater than a few billions of years became popular by the early 20th century. The invention of the mass spectrometer in the 1930s and the development of radioisotope dating techniques further pushed the estimate to over four billion years old. But how accurate are these dates?

Since the mid-20th century, scientists have used radioactive isotopes of uranium, thorium, rubidium, strontium, argon, and potassium to date rocks. Most of these methods require igneous rocks like lava and ash flows. Sedimentary rocks are usually mixtures of various weathered rock exposures, and therefore, give mixed ages if analyzed. Fossils, because they are found in sedimentary rocks, cannot normally be dated directly using radioactive isotopes, especially if thoroughly mineralized. Many people have gotten the wrong impression that paleontologists can stick a dinosaur bone into a machine and out comes the right age. This just does not happen in paleontology.

Determining absolute dates apart from historical records is very subjective and is laden with assumptions. First, it is assumed scientists can measure the decay rates of radioactive elements accurately. Many isotopes decay so slowly that the accuracy of these measurements is extremely difficult. Second, it's assumed that the decay rates never change, and have remained constant

Investigation of the physical properties of soils with radioisotope methods.

Becquerel and the Experiment with the Magnet

throughout all of time. But how do we really know the decay rate hasn't changed? Scientists at some of the world's leading institutions have reported variations in the decay rates of several elements over the course of a single year.[1] Creation scientists have also found evidence of accelerated decay of some isotopes in the past.[2] Neither of these first two assumptions seem as concrete as most secular scientists are led to believe.

Third, there is the assumption of a closed system, where elements and decay products are not supposed to be able to move in and out of the system over time. However, all rocks and minerals below the water table are in constant contact with water. Because water is the "universal solvent," it will selectively dissolve nearly every natural substance. Many dinosaur bones from Colorado are enriched in radioactive uranium because the porous interior of the bones allowed concentration via groundwater flow. Radioactive uranium ultimately decays into various isotopes of lead. If more uranium is selectively dissolved from the rocks compared to lead, it will look like a lot of uranium has converted to the lead isotope, resulting in age estimates that are much older than they should be, even millions or billions of years older.

The bottom line is that all radioactive isotope dating methods are preloaded with assumptions that cannot be truly verified. Secular scientists cannot go back in time and check the accuracy of their answers beyond a few thousand years at most, using tree rings and historical records. Unfortunately, they merely assume their results are factual, but they are presented as if unquestionable.

Secular geologists estimate absolute dates for most fossils by bracketing, as the rocks surrounding the fossil are often a mixture of various older, weathered materials. This makes any attempt at a single absolute date impossible to determine. The fossils themselves are often altered in some way also. So in order to "date" a fossil, a volcanic ash bed or a lava flow must be found above and below the layer containing the fossil. The absolute ages of the ash or lava are then estimated with radioactive dating techniques and assumed accurate. The fossil is then assigned an age between the two calculated dates, thereby bracketing the age of the fossil. More often than not, however, there will not be an ash layer of lava flow near the fossils. Then, the secular geologist must correlate the rocks and fossils with rocks and fossils elsewhere that have ash layers and/or lava flows nearby. Their methodology also includes the presupposition that the same fossils found anywhere else on earth are the exact same age. This is an unverifiable assumption, but it was used to assign dates to most fossils and to the geologic column.

It is far easier to accept the age of the earth described for us in the Bible than to rely on man's inaccurate methods. God was there as the only witness to His creation. Dinosaur fossils are not millions of years old as many scientists claim, but thousands of years old as we will see shortly. Many creation scientists have confirmed this young age through their research on comets, the earth's magnetic field, and other topics.

DINOSAUR SOFT TISSUE: CONCRETE EVIDENCE SUPPORTING A YOUNG AGE

Dinosaur bone is made of a mineral called calcium phosphate (hydroxylapatite), giving us the hard part of the bones, and a softer, organic part, the osteoid, principally collagen, produced by the bone cells, called osteocytes. Most fossilized bones replace the softer parts and the collagen with inorganic minerals from groundwater percolation. Previous studies had shown that collagen cannot last longer than 900,000 years without fully degrading, even under the best conditions. Therefore, it came as quite a shock to the scientific world when scientists first reported collagen from dinosaur bones. It even took an accident to make this discovery.

In her laboratory, Dr. Mary Schweitzer was dissolving some of the bone and mineral material from the center of a *T. rex* bone that was broken during extraction from the Hell Creek Formation in eastern Montana. She dissolved away all the hard bone parts by accident. Then she noticed some "sticky" material remained. The sticky material looked like bone soft tissue, collagen, blood vessels, and even red blood cells. She published her results in 2005[3] and was immediately attacked by critics claiming it could not be original material because collagen and other soft tissues cannot survive for 70 million years. She wondered how this could happen. It seemed impossible to her and her colleagues.

In an attempt to keep their old ages for dinosaurs, some secular critics tried to interpret these soft, pliable tissues as modern bacterial biofilms and not preserved original dinosaur tissue. However, in response, Dr. Schweitzer and her colleagues performed more studies on the soft tissue, publishing their results in subsequent papers.[4] They partially extracted collagen from the tyrannosaur leg bone and found a protein match of about 58 percent with bird collagen and 51 percent with frog and newt collagen. As further support that these are actual dinosaur soft tissues, another study examined 89 amino acids extracted from the aforementioned *T. rex* specimen, finding perfect matches with some modern animal proteins, thus, demonstrating the "impossible." The dinosaur soft tissue was real.

As mentioned above, chemical studies have shown that collagen cannot last longer than a million years before fully degrading, giving a maximum age to the dinosaur soft tissues. There are no known studies of how long red blood cells and blood vessels can last before fully degrading, but most likely we can expect a much shorter duration than collagen. Yet Dr. Schweitzer reported finding these types of soft tissues inside the dinosaur bones too.

Several other scientists have since reported finding soft tissues in other dinosaur bones, horns, and even in dinosaur embryos.[5] Organic preservation seems to be more common than ever

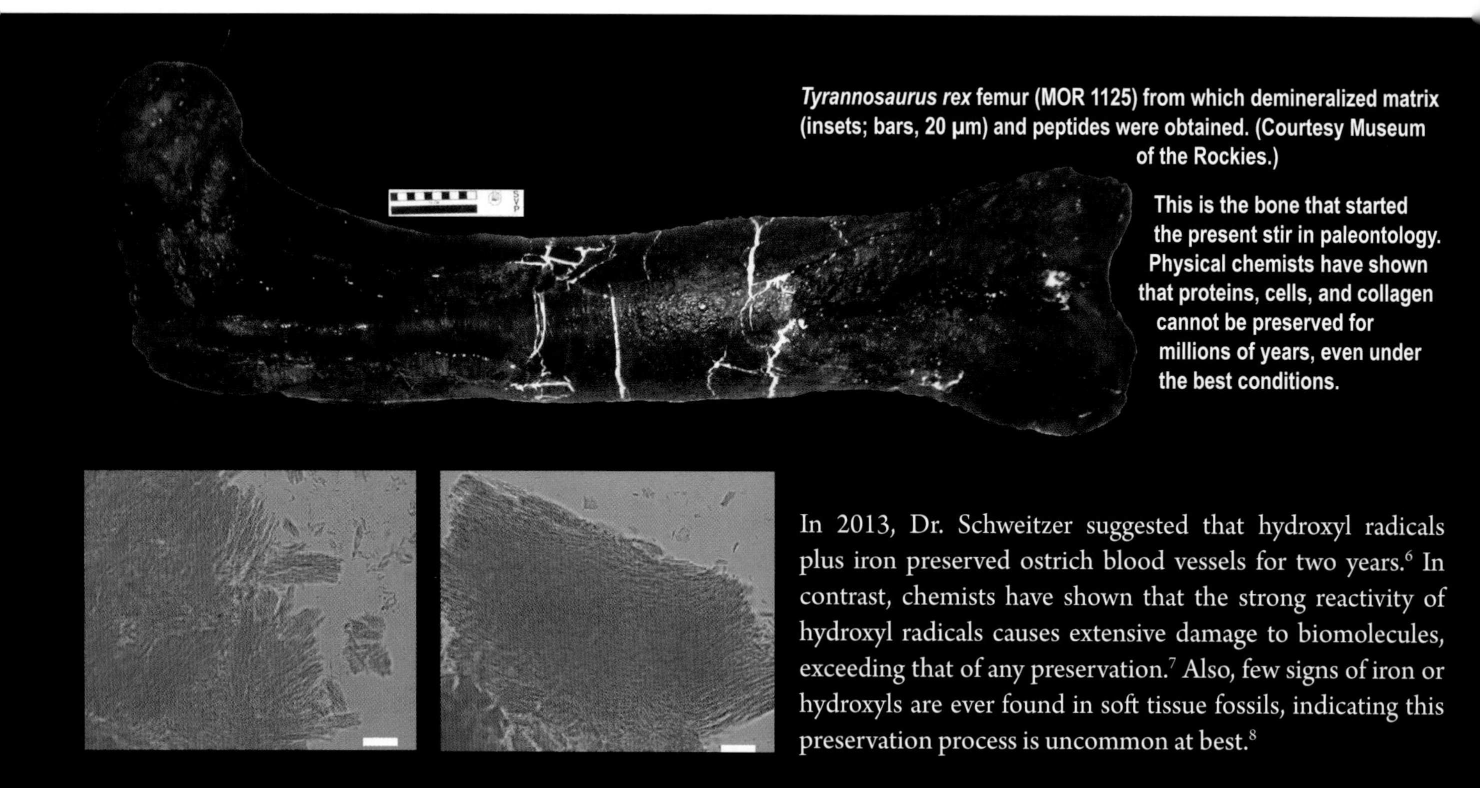

Tyrannosaurus rex **femur (MOR 1125) from which demineralized matrix (insets; bars, 20 μm) and peptides were obtained. (Courtesy Museum of the Rockies.)**

This is the bone that started the present stir in paleontology. Physical chemists have shown that proteins, cells, and collagen cannot be preserved for millions of years, even under the best conditions.

In 2013, Dr. Schweitzer suggested that hydroxyl radicals plus iron preserved ostrich blood vessels for two years.[6] In contrast, chemists have shown that the strong reactivity of hydroxyl radicals causes extensive damage to biomolecules, exceeding that of any preservation.[7] Also, few signs of iron or hydroxyls are ever found in soft tissue fossils, indicating this preservation process is uncommon at best.[8]

thought. Recent studies of organic compounds called melanosomes, that provide pigment and color, have also been found preserved, giving scientists a way to determine pigments of feathers of fossilized birds and even mosasaur skin (a swimming reptile).

Scores of peer-reviewed papers have reported some type of preserved, original, organic material in fossils purported to be many millions of years old, including reports of original organic preservation in fossils claimed to be over 400 million years old, and even over 500 million years old![9] Soft tissues in dinosaurs have been found in tyrannosaurs, hadrosaurs, titanosaurs, psittacosaurs, ceratopsians, and others. Discoveries of preserved dinosaur soft tissue are no longer unusual.

The finding of actual soft tissue in fossils that are supposedly millions of years old has evolutionists scratching their heads to explain. They are scrambling, trying to come up with "miracles" of preservation. Many secular scientists cannot fathom that dinosaurs are only thousands of years old because they "walk in the futility of their mind, having their understanding darkened, being alienated from the life of God, because of the ignorance that is in them, because of the blindness of their heart" (Eph. 4: 17–18). Creationists are often criticized for relying on a literal interpretation of the Book of Genesis, but it is the evolutionists who are denying factual data. They are relying on their beliefs in old ages in spite of the evidence. Evolutionists cannot admit this can only point to a young age for the dinosaur bones, and the dinosaurs themselves. The case for thousands and not millions of years is stronger than ever.

Detail of the head of *Scipionyx samniticus* in the Natural History Museum in Milan, Italy. The fossil preserves in an exceptional way clear traces of soft tissue. Picture by Giovanni Dall'Orto, April 22, 2007.

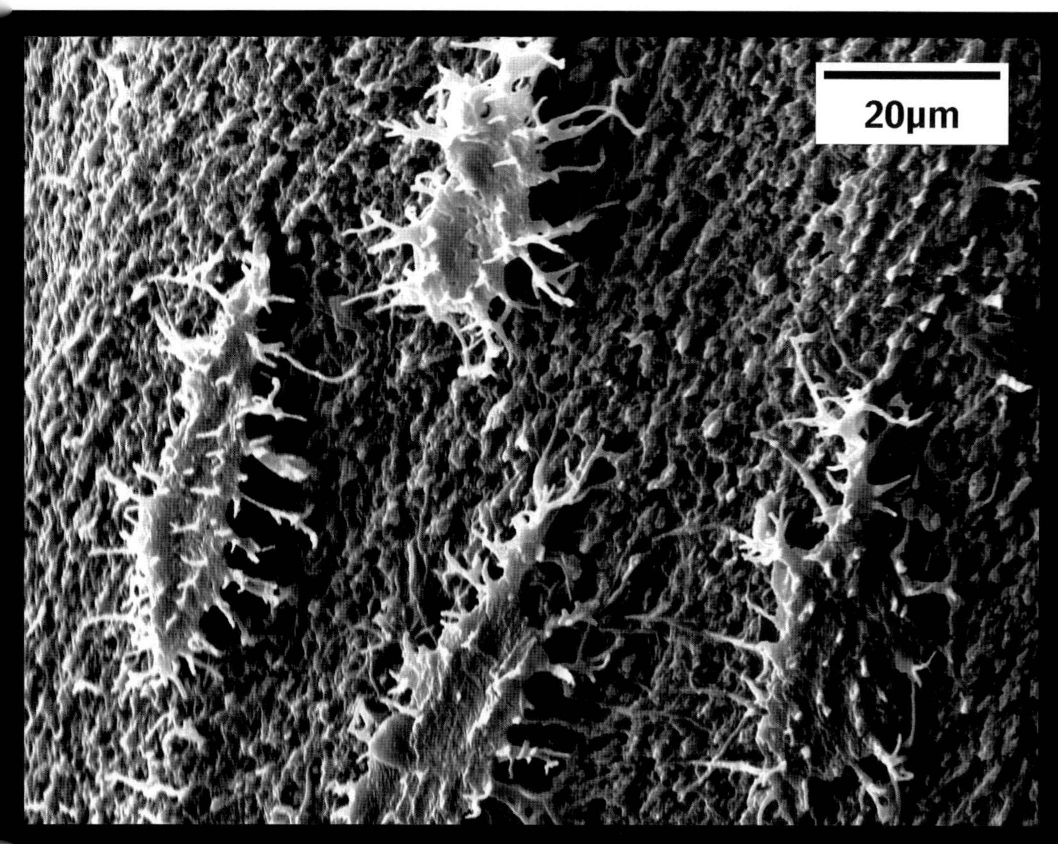

Scanning electron microscope image of decalcified bone material. Image shows intact osteocyte cells lying on fibrillar bone matrix from the interior of a *Triceratops horridus* horn. Tiny filipodial processes are visible anchoring the cells to the bone matrix. The supraorbital horn specimen was found in the Hell Creek Formation near Glendive, MT. Image taken from the *Creation Research Society Quarterly* (2015, vol. 51, p. 255). Used by permission.

CARBON-14 IN DINOSAUR FOSSILS SHOWS A YOUNG AGE

Why haven't the dinosaur soft tissues and organic proteins that have been found recently been tested for carbon-14? It is because of evolutionary dogma, not because of any scientific data.

Dinosaur fossils are thought to be so old that there should be no measurable C-14 left even if original material were preserved. Anything truly older than 100,000 years should have no measurable C-14, as all of it should have decayed to undetectable levels. But what do we see when we test dinosaur fossils?

Recall, absolute dates are based on the decay rate of unstable (radioactive) isotopes. Decay rates are converted to a half-life, which is the time it takes for one-half of the radioactive element to decay to another element. Carbon-14 makes a good example to illustrate radioactive decay. Carbon exists as three isotopes — C-12, the most common variety; C-13, the next most common variety; and C-14, the rarest and only radioactive variety. All three carbon isotopes have six protons which define them as carbon. Carbon-12 has six neutrons, C-13 has seven neutrons, and C-14 has 8 neutrons, making the latter unstable. Carbon-12 and C-13 remain C-12 and C-13 forever, respectively. Carbon-14, by contrast, spontaneously decays into nitrogen-14 with time. Carbon-14 has a half-life of about 5,730 years at present, meaning that one-half of any quantity of C-14 will convert to N-14 in that time. By determining how much C-14 is left in a sample, we can estimate its age. If we have one-quarter of what we started with, we can assume two half-lives have passed, and therefore determine that the sample is 11,460 years old (5,730 + 5,730).

The Institute for Creation Research (ICR) sent a small piece of fossil paddlefish that they had in their collection to an independent laboratory for C-14 analysis. This paddlefish was found in the Eocene Green River Formation of Wyoming, supposedly about 50 million years old. There should have been no detectable C-14 remaining if the fossil were truly millions of years old. The lab found enough C-14 to date the fossil at 33,530+170 years before present (BP). ICR scientists still believe this estimated age is too old for the biblical timeframe, and this matter is explained below.

ICR also sent two samples from a Cretaceous-age hadrosaur dinosaur to the same independent laboratory for C-14 analysis. These samples were from the Hell Creek Formation in Montana, supposedly about 70 million years old, and should also have contained no detectable C-14. The lab found enough C-14 in each sample to determine dates of 28,790+100 years BP and 20,850+90 years BP.

In addition, ICR acquired dinosaur bone samples from a *Stegosaurus* and a *Diplodocus* from the Morrison Formation in Utah, supposedly 150 million years

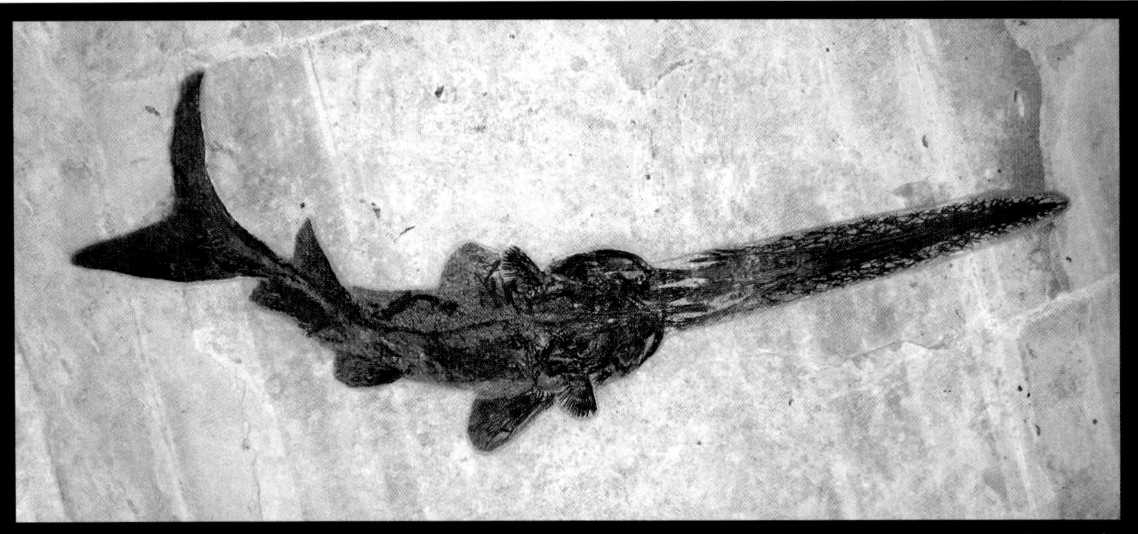

Fossil paddlefish from the Green River Formation, Wyoming. It is identical to paddlefish living today, demonstrating stasis of the created kind and not evolution.

old. These samples also contained enough C-14 to be dated at 21,830+70 years BP for the stegosaur and 26,890+90 years BP for the diplodocid.

Over 20 other dinosaur bone samples and egg shell fragments were also tested for C-14 at an independent laboratory. These results also came back in the range of 20,000 to 50,000 years BP. What is going on? Obviously, the C-14 dates are consistently showing dates in the thousands and not in the millions of years range. There is substantial detectible C-14 in many dinosaur fossils. Is this all merely contamination, as secular scientists claim? ICR scientists interpret these results as strong evidence that dinosaurs are not millions of years old. Critics may argue these C-14 dates are all due to contamination, but great care was taken to sample the bones, using bone material from the inside, not the edges. The commercial lab took great care not to introduce contamination, and the consistency in the C-14 dates from samples taken across three states, and in three vastly different rock layers, implies there was no contamination.

During the C-14 analyses, ratios of a stable isotope of carbon, carbon-13, were also recorded for these samples and reported as the standardized ratio δ13C. This variety of carbon, although rare, does not spontaneously decay. However, values of δ13C can help demonstrate if contamination has occurred prior to sampling. Usually, the δ13C ratio in the mineral part of any bone, called bioapatite, is significantly different from the ratios of δ13C in the organic part (collagen) of the bone. We therefore expect actual preserved fossil soft tissue and fossil bone to reflect these differences in δ13C values, if they are real, and not contaminated in some way after burial in the rock.

What did the ICR scientists and the others find? Most of the bioapatite δ13C ratios fell in the expected ranges for carbonate minerals derived from uncontaminated bone. And the δ13C ratios for the organic collagen showed significantly different δ13C ratios, close to what was expected for organic material in uncontaminated samples. The measured δ13C ratio differences between the bioapatite and in the extracted organic components strongly argue against contamination.

The C-13 measurements provided an independent check on the validity of the C-14 results reported above. If everything was uniformly contaminated, all δ13C ratios should leave nearly the same values. These findings indicate the measurable amount of C-14 found in the bones is real, and demonstrate a young age for the dinosaurs, just as the Bible says!

Why 20,000 to 30,000 years old and not 4,000 to 5,000 years old? Unfortunately, C-14 dating is based on the same assumptions of all radioactive methods, in particular, the modern ratio of C-14/C-12. We know that C-14/C-12 ratios in the atmosphere have not stayed constant even in the past few thousand years because corrections had to be made to the C-14 dates to match tree ring data.

Creation scientists point out that the pre-Flood atmosphere was likely much higher in carbon dioxide (CO_2), possibly even 500 times higher than today's atmosphere. This estimate is based on the amount of coal, oil, and other organic debris buried in the sediments of the Flood.[10] A higher level of CO_2 would have greatly altered the pre-Flood C-14/C-12 ratio, assuming the cosmic rays that produce C-14 were not much different from today. Increasing the atmospheric C-12 content would have made the starting C-14/C-12 ratio much smaller in all the fossils buried in the Flood, making them seem older from the start. When the C-14 in the buried fossils is measured and compared to today's C-14/C-12 ratio, it would result in inflated "ages" of up to ten times their true age. So, instead of 4,500 years old, wrongly applying the modern ratio of C-14/C-12 gives an assigned age of 45,000 years. Unfortunately, we only have the modern ratio of C-14/C-12 to work with, so these "true" ages will always be inflated by about one order of magnitude. But, regardless of which starting ratio is used, the presence of detectable C-14 in ancient fossils demonstrates that dinosaurs are not millions of years old!

1

$${}^{14}_{7}N + {}^{1}_{0}n \rightarrow {}^{14}_{6}C + {}^{1}_{1}p$$

2

$${}^{14}_{6}C \rightarrow {}^{14}_{7}N + {}^{0}_{-1}e + \overline{\nu}$$

Display of the bones of the *Hadrosaurus* at the Philadelphia Academy of Natural Sciences. 35 bones of the *Hadrosaurus* were dug up in the 1800s and formed the first "fairly complete dinosaur skeleton." This display shows which of the bones were found.

CHAPTER 4: DINOSAUR DISCOVERIES

THE VERY FIRST DINOSAUR DISCOVERIES

What was likely the first written description of a dinosaur? Around 300 B.C., Chinese scholar Chang Qu wrote about "dragon" bones found in what is now Sichuan Province, China. He would have likely called them dinosaurs today, and he would have received credit for their first discovery. In the Book of Job, written possibly as early as 2000 B.C., we read the description of the Behemoth, which seems to reveal a perfect image of a living sauropod dinosaur, or some large, similar-sized dinosaur. This biblical text, therefore, seems to be the first and oldest written record of a dinosaur. Unfortunately, most secular scientists and secular historians scoff at this and refuse to take this seriously, as they believe that men and dinosaurs never coexisted. But for Bible-believing scientists and scholars, the evidence is more compelling.

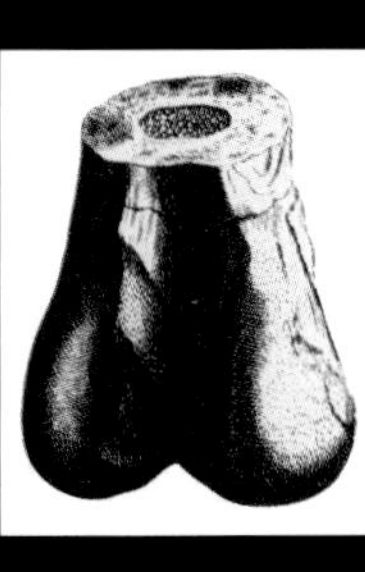

The first drawing and modern description of a dinosaur bone in the scientific literature was by Oxford University clergyman and chemist Robert Plot in 1677. He believed the specimen he had examined, found near Oxfordshire, was from a modern elephant brought to Britain by the Romans, and not from an extinct creature at all. Paleontologists now believe the original specimen, although lost, was part of the femur of a *Megalosaurus*, a theropod dinosaur found in Jurassic strata in Britain.

Nearly 150 years later, *Megalosaurus* also became the first "dinosaur" to be named. It was first described by the Reverend William Buckland, Reader of Mineralogy at Oxford University. Apparently, Buckland obtained several teeth and a jaw fragment around 1815 from an area near Stonesfield, Oxfordshire, England. He showed the specimens to numerous people, including the French anatomist Georges Cuvier, before publishing his formal name and description in 1824, using the same name used for the dinosaur by English surgeon James Parkinson in 1822, but who failed to provide a description.

Robert Plot
1640–1696

American scientist Dr. Joseph Leidy, professor of anatomy at the University of Pennsylvania, described the first dinosaur found in the United States. Dr. Leidy named the *Hadrosaurus* in 1858 from a collection of bones found by W.P. Foulke near Haddonfield, New Jersey. Just a few years earlier, in 1856, Dr. Leidy had identified a few teeth from Montana as those of a dinosaur, and the first discovered in the Western Hemisphere. British sculptor Benjamin Waterhouse Hawkins mounted the bones of the *Hadrosaurus* in 1868 at the Academy of Natural Sciences in Philadelphia. Hawkins didn't quite get the posture of the dinosaur correct, but that was not realized until about 100 years later.

The first discovery of a dinosaur footprint was by Pliny Moody in 1802 while plowing a field in New England. Because dinosaurs were not known at the time, the significance of his discovery was largely unnoticed until much later. Some of the earliest studies of these dinosaur footprints were done by Edward Hitchcock in the mid-19th century. Unfortunately, most scientists of the time were caught up in the search for dinosaur bones, and footprints were neglected until very recently.

Joseph Leidy
1823–1891

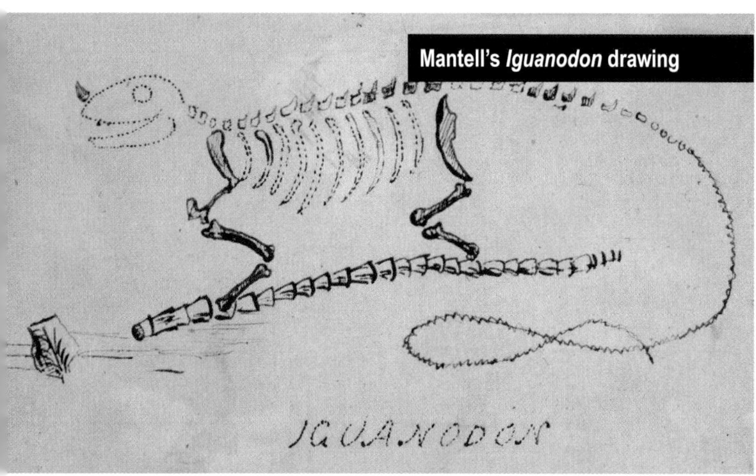

Mantell's *Iguanodon* drawing

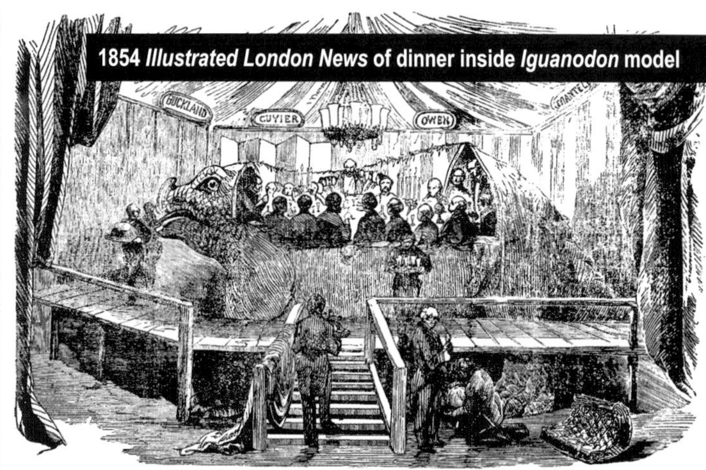

1854 *Illustrated London News* of dinner inside *Iguanodon* model

Iguanodon footprints

Modern reconstructions of *Iguanodon* show the animal walking on all fours, and the straight tail held high in the air. Footprint evidence supports this modern version, showing large back footprints, smaller front footprints, and no tail marks.

THE CHANGING IMAGE OF *IGUANODON*

The first herbivorous dinosaur was named by a contemporary of Buckland, physician Gideon Mantell. He and his wife, Mary Ann, reportedly found a few teeth and bones in 1822 from which he named *Iguanodon* in 1825. The story goes that Mary Ann went along with her husband, Gideon, on a house call near Cuckfield in Sussex, England, in 1822. While there, she found some fossil teeth in a pile of rocks set aside for road repair. Confused as to what they belonged to, Gideon Mantell sent the teeth and other bone pieces to Georges Cuvier in Paris. He incorrectly identified them as rhinoceros teeth, and another strange-looking bone he thought was a rhinoceros horn (found later to be a thumb spike).

After hearing a talk by William Buckland on his discovery of the reptile *Megalosaurus* in 1824, Mantell started thinking that the teeth and bones his wife found may have been from an herbivorous reptile. He examined a modern iguana skeleton and noticed they were remarkably similar; his specimen was just bigger. He published his conclusions in 1825, naming *Iguanodon* (iguana tooth), initially thinking the *Iguanodon* was 60 feet long.

Dr. Gideon Mantell drew this rendition of the dinosaur *Iguanodon* based on the similarity of the teeth to a modern iguana. He put the thumb spike on the nose by mistake, as he only found a few bones and didn't really have anything for comparison.

Mantell also found fragments of three other extinct reptiles (later identified as dinosaurs) in 1833. Unfortunately, most of the dinosaurs found and named in England in the 19th century were fragmentary. Even today, there are few articulated specimens found in England. It is, therefore, all the more amazing that Sir Richard Owen was able to define an entire extinct group of reptiles from such incomplete evidence.

In 1854, artist Benjamin Waterhouse Hawkins created life-sized models of several of the extinct reptiles discovered in England, including the *Megalosaurus*, *Hylaeosaurus*, and *Iguanodon*. These statues were part of the grand reopening of the Crystal Palace exhibit in Sydenham, England. Hawkins placed the thumb spike on the snout of his *Iguanodon* model because he and Sir Richard Owen believed that was where it belonged.

Sir Richard Owen hosted a dinner inside the still unfinished model of Hawkins'

Lithograph with later hand coloring combining two of Benjamin Waterhouse Hawkins' six wall posters "Waterhouse Hawkins Diagrams of the Extinct Animals" produced for the Department of Science and Art in 1862. It includes nearly all the dinosaurs known up to that date. From left, two *Iguanodon* sculptures (with "thumb" spike mistakenly placed on nose), *Hylaeosaurus* in the middle, and to the right, two *Megalosaurus* sculptures. On cliffs in background are *pterosaurs* (not dinosaurs). In 1855 Waterhouse Hawkins produced life-sized reconstructions, with the scientific help of Richard Owen, for the gardens of the Crystal Palace at Sydenham in London, England.

Iguanodon in 1854. Most of the prominent scientists of the day dined inside the four-legged reptile's hollow interior, complete with thumb spike on the nose. It wasn't until 1878, with the discovery of many complete, articulated *Iguanodon* specimens in Bernissart, Belgium, that the confusion over where to correctly place the thumb spike was finally resolved.

The discovery in Bernissart of many complete specimens made it possible to reconstruct accurate, full-mounted skeletons. The fossils were found over 1,000 feet below the surface in a coal mine. Because the bones were filled with pyrite (fool's gold), some of the miners wrongfully thought they had found tree trunks of solid gold. The Belgium discovery yielded 40 partial and/or complete *Iguanodon* individuals, many other reptiles, 3,000 fish, and 4,000 plant fossils. The odd mix of fish and dinosaurs doesn't surprise creation scientists, as the Great Flood commonly mixed fish and dinosaurs in its catastrophic deposits.

The first reconstructions of the *Iguanodon* specimens at Bernissart fell into the capable hands of Louis Dollo. He was born in Lille, France, but came to Brussels as an engineer. Hired by the Royal Museum in Brussels to reconstruct the *Iguanodon* specimens, he began breaking new ground in the scientific world by creating many new skeletal reconstructions for public view. He had a distinct advantage over his predecessors because he was able to use complete dinosaurs to assist him in this endeavor. However, complete specimens buried rapidly and distorted in the Flood are not the same as visualizing a standing, life rendition. He needed a starting point, an example to follow. Dollo chose the modern skeletons of an emu and a wallaby for reference. He used an iron framework to make the dinosaurs stand and appear as they would in "real life." Unfortunately, he forced the dinosaurs to stand in a three-point stance, with tail dragging, even dislocating some bones in the process.

Dollo's dinosaur reconstructions established the model for all subsequent dinosaur positions and overall posture for the next century. It wasn't until recently (1980s) that we changed the posture of dinosaurs to better reflect their footprint evidence and better illustrate a sense of balance. Today's modern reconstructions show dinosaurs with tails held high, not dragging, and much thinner and agile than renditions in the past. Dollo did groundbreaking work, stepping forward into unknown territory, but additional findings have slowly caused correction to some of his early misconceptions.

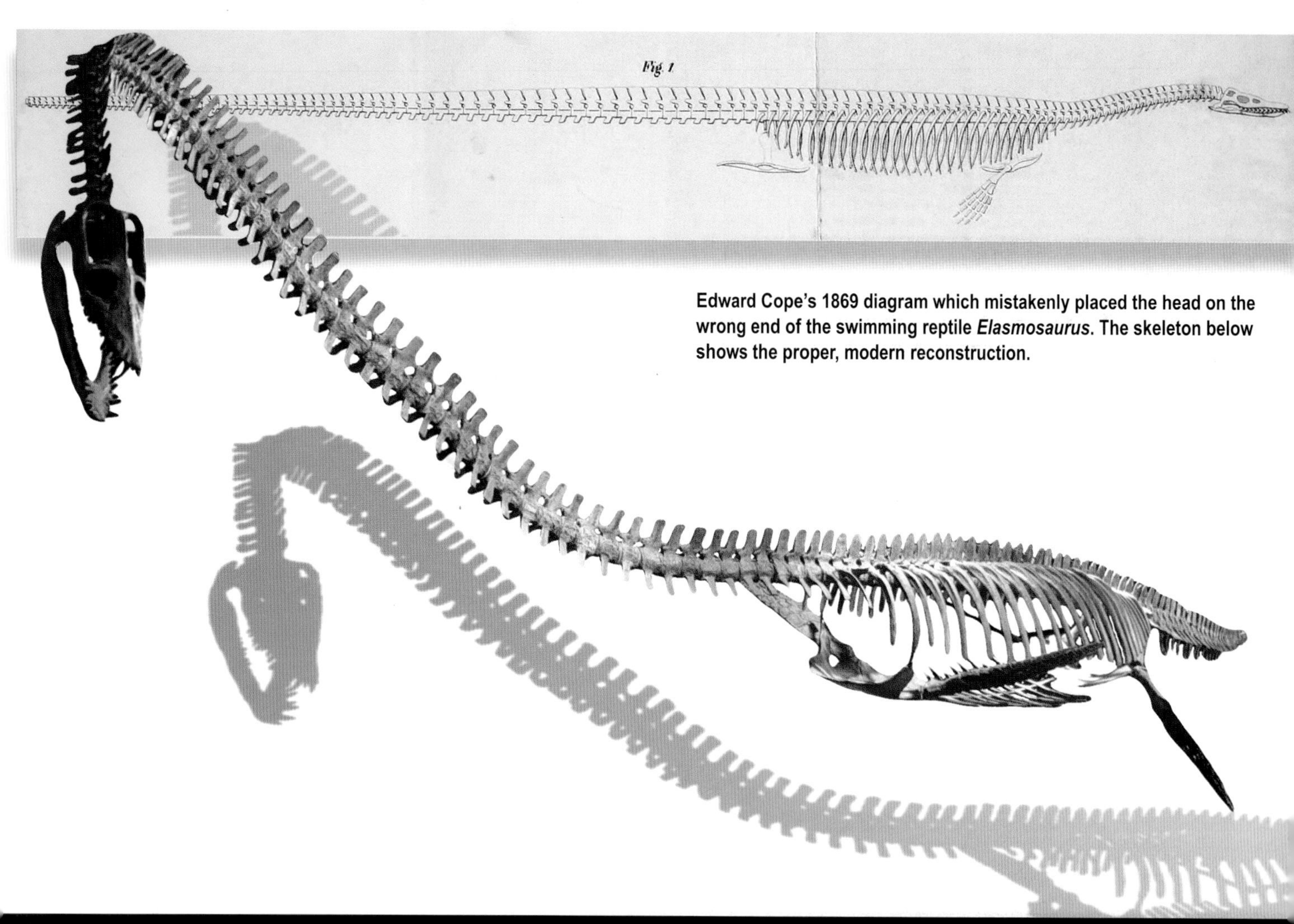

Edward Cope's 1869 diagram which mistakenly placed the head on the wrong end of the swimming reptile *Elasmosaurus*. The skeleton below shows the proper, modern reconstruction.

THE GREAT DINOSAUR RUSH

The great Dinosaur Rush of the Gilded Age (1870s–1890s) was fueled by competition between two rival scientists, Edward Drinker Cope and Othniel Charles Marsh. Cope (1840–1897) was from a wealthy Philadelphia family who had made a substantial fortune in shipping. Marsh (1831–1899), whose uncle was the millionaire banker George Peabody, was from New York. Although Cope never earned any formal college degrees (he was awarded an honorary master's degree), Marsh graduated from Yale and obtained a master's degree from Sheffield Scientific School in 1862. The two first met in Berlin in 1863.

In 1869, E.D. Cope invited O.C. Marsh to view his recently constructed skeleton of *Elasmosaurus platyrus* at the Academy of Natural Sciences in Philadelphia. The 35-foot swimming reptile was only the second reconstructed skeleton of an ancient reptile in the United States, after the dinosaur *Hadrosaurus*. Marsh quickly ascertained that Cope had put the head on the tail, and the feud began. They called in Dr. Joseph Leidy, a medical doctor who was considered America's expert in paleontology, who confirmed that Cope had indeed made a mistake. Cope quickly tried to buy back all the journal articles detailing his "mistake," but his reputation was permanently damaged.

Following their infamous disagreement over the reconstruction of Cope's *Elasmosaurus* in 1869, the pair squared off for over 30 years of intense competition, draining their respective financial resources in the process. The battle for the most new species was on. Both Cope and Marsh often hired competing crews of dinosaur hunters to scour the American West, searching for more and newer finds. Some crews would send a few samples back to each scientist, and would later submit the rest of the bones to the highest bidder, which was usually Marsh. In order to better preserve their prize bones, the crews developed some of the excavation techniques still in use today by modern dinosaur hunters.

Unfortunately, in their haste and in their fierce competition, some quarries were destroyed by rival crews, and mistakes were often made in naming new species. Many famous dinosaur quarries were opened during the "Bone Wars," including ones near Como Bluff, Wyoming; Morrison, Colorado; and Cañon City, Colorado. This feud introduced the world

Edward Drinker Cope
1840–1897

Othniel Charles Marsh
1831–1899

to some of the most famous dinosaurs such as *Apatosaurus*, *Stegosaurus*, and *Triceratops*. In the end, Marsh also helped establish some of the foundations of the dinosaur classification system still in use, including the suborders of sauropods, theropods, and ornithopods.

E.D. Cope was constantly thwarted by Marsh ever since his mistake in 1869. He was always a day late and a dollar short, as the saying goes. He did publish over 1,400 papers in his lifetime, still a scientific record. In retrospect, many of those papers were self-published and at his own expense. But he named many other species besides dinosaurs, including, fish, lizards, and mammals. Later in life, Cope had to sell much of his fossil collection, at a fraction of his costs, to make ends meet. Cope was also convinced, as was the theory of the day, that his brain was bigger than Marsh's. Upon his death, Cope's brain was weighed and measured, but alas, when Marsh passed away two years later he left strict orders to make sure his skull remained intact.

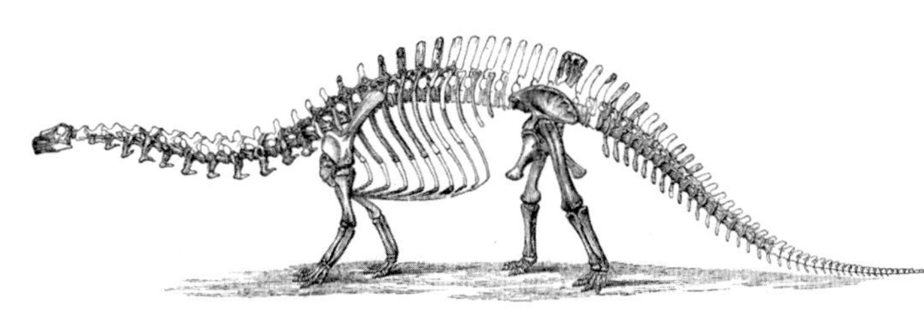

Illustration of reconstructed *Brontosaurus* skeleton by O.C. Marsh with the wrong head.

Professor Marsh convinced his Uncle George Peabody to make a large donation to Yale College in 1863 to establish the Peabody Museum of Natural History. He also used this donation to gain favor with the college. In 1866, he was named the chair of paleontology at Yale College, funded again by his Uncle George. O.C. Marsh still holds the record for most dinosaur genera named and discovered.

Othniel Charles Marsh, *American Paleontologist* Cartoon from "Punch" (London, September 13, 1890)

Unfortunately, Marsh also was prone to making a few mistakes. He named the *Brontosaurus* in 1879, but like most sauropods, his specimen had no head. Modern paleontologists think the head on sauropod necks was only weakly attached and that the heads easily became separated during catastrophic burial. Marsh later placed a head that seemed to fit on his reconstructed skeleton but had been miles away. Much later, paleontologists recognized this was the wrong head. It was from a similar-sized *Camarasaurus*. Unfortunately, it took nearly 80 years for all the correct heads to be replaced on the numerous museum specimens of *Brontosaurus* scattered across the globe.

Prof. Marsh also named the genus *Triceratops*, using a skull discovered by John Bell Hatcher in 1888 from the Judith River Formation in Montana. Marsh initially ignored the first *Triceratops* skulls that he came across, confusing their well-preserved remains with those of modern bison found scattered across the same rock exposures. Although found in Upper Cretaceous rocks (conventionally dated at 70 million years ago), Marsh thought the skulls looked very young, just as creationists correctly continue to point out. In fact, creation scientists have found soft tissues preserved in a *Triceratops* horn and a frill.

OSBORN, BROWN, AND STERNBERG: THE GREATEST GENERATION OF DINOSAUR HUNTERS

After Cope and Marsh came the next generation of dinosaur hunters. These included Henry Fairfield Osborn, Barnum Brown, and Charles Sternberg. Osborn (1857–1935) started the American Museum of Natural History's (New York) paleontology department in 1897. The museum had previously purchased E.D. Cope's fossil collection for a paltry $32,000, and Osborn was determined to make the American Museum a major home to dinosaurs. He hired Barnum Brown, Walter Granger, and eventually, Charles Sternberg to collect for his museum. Osborn's employees dug right in at Como Bluff, Wyoming, finding 517 specimens in seven years, including many *Apatosaurus*, *Camarasaurus*, and *Diplodocus* specimens.

Henry Fairfield Osborn had a "dark" side as well. He was a big proponent of the unbiblical philosophy of eugenics, convening a conference on the subject in 1921. He was also a staunch believer in evolution and held that the very character of humans could be determined by heredity.

Barnum Brown was born and raised in Kansas, and was named after Phineas T. Barnum, following a family visit to the circus. He began collecting fossils as a child and continued his collecting as a student at the University of Kansas. It was on a summer dig in Wyoming in 1895 that he found his first dinosaur, a *Triceratops* skull. He was subsequently hired by the American Museum of Natural History during the summer of 1897, while on a dig site in Wyoming. Brown quickly became Prof. Osborn's favorite dinosaur hunter.

Barnum Brown found the first specimen of *Tyrannosaurus rex* in 1902, while excavating for the American Museum of Natural History in Montana. Professor Osborn named and described the specimen in 1905. Brown then went on to find a second *T. rex* skeleton in 1908. By the time his dinosaur-digging days were over, Barnum Brown had become one of the most prolific dinosaur hunters of all time. He continued to collect for the American Museum thoughout his career. He was even brought in by the National Park Service as an advisor at Dinosaur National Monument in 1937.

Charles Sternberg began his fossil-collecting with the backing of E.D. Cope in the 1870s. Although from Kansas, Sternberg was asked by Cope to spend the summer of 1876 digging with him in Montana. It was that same year that Cope and Sternberg developed the use of plaster of Paris and burlap strips to encase the bones and protect them for shipment.

Tyrannosaurus rex **skeleton (the specimen AMNH 5027) at American Museum of Natural History.**

American Museum of Natural History

Sternberg worked for the American Museum for only part of his career. He and his three sons scoured the American West and Canada as freelance collectors, even working for the British Museum. He found one of the first "mummified dinosaurs," an *Edmontosaurus* (duck-bill) with remarkable preservation of skin and abdomen, including the stomach contents of pine needles, twigs, seeds, and roots.

Sternberg and his youngest son, Levi, found and excavated a nearly complete, 30-foot, duck-billed dinosaur, including skin imprints, from western Canada in 1916. They sold it to the British Museum but were never paid, as the ship and dinosaur, along with a second specimen, were sunk by a German U-boat in transit.

DINOSAUR QUARRIES OF THE MORRISON FORMATION

Numerous quarries from the Upper Jurassic Morrison Formation were opened by dinosaur hunters during the 1870s–1890s, the period of the Great Dinosaur Rush. Most of the crews were working for either E.D. Cope or O.C. Marsh, who funded and fueled the search. The Morrison Formation is a rock unit that has properties that are distinctive enough to be mapped and identified across parts of many western states. And it is one of the few rock units to keep the same name from state to state. It is more common for a rock unit to change names across state lines and just as common to change names from place to place within a single state. The Morrison name was derived from near where the rocks were first described, outside of the little town of Morrison, Colorado. Morrison is just south of Dinosaur Ridge, along the Dakota Hogback of the Front Range, west of Denver. The first discovery of dinosaur bones at the Morrison site was made by Arthur Lakes and Captain Henry Beckwith, while surveying the tilted strata on March 26, 1877. Although Beckwith initially thought the bones were trees, Lakes quickly realized they were bone fossils. He sent samples to both Marsh and Cope to see who was most interested, and who would pay for their excavation. Marsh wrote back first, offering $100 a month to begin quarrying. Lakes opened the Morrison quarry in 1877, and the discovery of *Apatosaurus* and *Stegosaurus* followed shortly thereafter.

The Morrison Formation consists of distinctive red, purple, gray, and brown-colored shale, mudrock, and sandstone. It extends across a large portion of the American West, and contains most of the same rock types and order from Utah to Wyoming to Colorado. Numerous bone beds containing thousands of catastrophically deposited dinosaur bones have been found and quarried from this unit. Creation scientists explain the Morrison Formation as a rapid depositional event that spread across these states during the great Flood. The dinosaurs were trapped and buried as part of that event.

Bone Cabin at Como Bluff, WY

The dinosaurs at Como Bluff, Wyoming, were discovered in the Morrison Formation by Union Pacific Railroad employee William Harlow Reed in 1877. He and his colleague, William Edwards Carlin, wrote to Marsh and were soon put on the payroll digging dinosaurs. They found *Camarasaurus, Stegosaurus,* and *Brontosaurus* (later becoming *Apatosaurus*), as well as fish, frogs, lizards, turtles, crocodiles, pterodactyls, and mammal fossils. There were so many dinosaur bones at Como Bluff that a sheepherder had used the bones to construct a cabin made of bone, the so-called Bone Cabin, made of 26,000 dinosaur fossils.

The Morrison Formation quarry at Cañon City, Colorado, (near Royal Gorge), was opened in 1877 after being discovered by O.W. Lucas, the local school superintendent. Lucas sent some bones to Cope, who instructed Lucas to hire a crew and begin digging. They discovered *Camarasaurus* at this site. At a nearby site, Marsh's crew, led by Samuel Wendell Williston, found the remains of another large sauropod, dubbed *Diplodocus* by Marsh in 1878.

Earl Douglass, an employee of the Carnegie Museum in Pittsburgh, was searching for bones south of the Uinta Mountains, near Vernal, Utah, in 1908 when he spotted a *Diplodocus* leg bone at the bottom of a ravine. He wasn't able to return to the site until a year later to continue his search.

Barnum Brown and Henry Fairfield Osborn at Como Bluff quarry, Wyoming, during the 1897 expedition by the American Museum of Natural History, New York.

***Apatosaurus* (formerly *Brontosaurus*) *excelsus* in the Yale Peabody Museum of Natural History.**

Discoverer William H. Reed 1879

Locality Como Bluff, Wyoming

Viewing dinosaur tracks in Wyoming.

What he found was well worth the wait. Douglass initially discovered the articulated bones of a sauropod that was named *Apatosaurus louisae*, in honor of Mrs. Andrew Carnegie, wife of the museum's benefactor. He also found the remains of over 20 other large dinosaurs and the bones from nearly 300 additional individuals. Douglass struggled trying to comprehend why these dinosaur skeletons were jumbled all together in one place, creating one of the largest collections of Morrison Formation dinosaurs ever discovered. He could only envision a huge river flood, but creation scientists realize this was a deposit from the great Flood. Eventually this site was designated as Dinosaur National Monument in honor of the great discoveries made by Douglass and his crew.

The bones of at least 70 individuals, including 44 *Allosaurus* specimens, were found in the Morrison Formation at Cleveland-Lloyd Quarry, near Price, Utah. Oddly, nearly three-fourths of the 12,000 bones found in this quarry were from predatory dinosaurs. The Morrison Formation has also yielded over 4,000 bones from a single site at Howe Quarry, near Shell, Wyoming, and over 1,000 bones at a site near Thermopolis, Wyoming. Like Douglass, today's evolutionists continue to struggle to explain these mass death assemblages. They are hindered by their belief in uniformitarianism to using large river flood events to justify the catastrophes. Creationist scientists, however, explain each site within the Morrison Formation as evidence of the great Flood, when ten thousands of dinosaurs were simultaneously and catastrophically buried in water and sediment across the American West.

Brachiosaurus altithorax sauropod dinosaur from the Jurassic of Colorado, USA (mount owned by Field Museum of Natural History and on display at O'Hare Airport's Terminal B in Chicago, Illinois, USA).

THE GREAT EAST AFRICAN DINOSAUR EXPEDITION

Between 1909 and 1912, the Germans mounted one of the most spectacular dinosaur expeditions of all time. The site was located near the village of Tendaguru in German East Africa, now part of the country of Tanzania. The director of the Berlin Museum, Dr. W. Branca, was shown some fossils collected by paleontologist Eberhard Fraas near the site in 1908. Branca immediately organized an expedition to collect more fossils. Paleontologists Werner Janensch and Edward Hennig were placed in charge. The expedition began in 1909 with 170 local laborers, but by 1912 had blossomed to 500. The total at the site in 1912, including the scientists, the families of the laborers, and supporting personnel, was estimated at 700 to 900 people. About 250 tons of bones and rock were excavated from the site and hand-carried on a four-day journey to the coast for shipment to Germany.

One of the most spectacular finds made by the expedition was a nearly complete *Brachiosaurus*. This specimen still resides as the showpiece of the Berlin Museum, surviving the bombings of World War II. This discovery also demonstrated that some dinosaurs had greater distribution than previously thought, as the first, partial skeleton of *Brachiosaurus* was unearthed in the Western United States and described by Elmer Riggs in 1903. Janensch described the German specimen in a publication in 1914. Oddly, no one at the time offered an explanation of how the same dinosaurs could have coexisted on separate continents.

Today, some paleontologists consider the Berlin specimen to be a separate, but very similar, species to *Brachiosaurus*, and have renamed it *Giraffatitan brancai*. Unfortunately, there is no consensus on the name, and many other scientists still use the original name.

Famous for its sneeze in the *Jurassic Park* movie, the *Brachiosaurus* is considered by many to be the heaviest dinosaur found in the United States, weighing in between 30 and 55 metric tons. It may be outweighed by *Argentinosaurus* from South America, estimated at 70 metric tons, but this specimen is very incomplete. The newest contender for the heaviest dinosaur seems to be *Dreadnoughtus schrani*. This huge titanosaur is about 70 percent complete and was estimated at 59.3 metric tons.

Brachiosaurus differed from most sauropods (long-necks) because it had longer front legs than back legs. This gave the dinosaur the high, upward-angled neck and the elevated head that are its distinguishing features. Creationists argue that the presence of similar brachiosaurid fossils in Africa and North America should not come as a surprise. The supercontinent that existed before the great Flood would have provided physical access to both areas, making it easy to have deposited the remains of brachiosaurids on what are now separate continents.

Photos from the 1909–1911 Tendaguru expedition to German East Africa (now Tanzania). Bottom right shows German paleontologist Werner Janensch.

THE CENTRAL ASIATIC EXPEDITIONS

Five separate expeditions journeyed into Central Asia to search for fossils between 1922 and 1930. The goal of each of the missions was to unlock the fossil history of heretofore unknown territory. Evolutionist Henry F. Osborn, president of the American Museum of Natural History at the time, believed that Asia was the origin of many dominant modern land animals, including humans. Scientists had previously gathered extensive fossil collections in the Americas and in Europe, but information in Asia, between these two areas, was lacking, leaving a gap in knowledge that Osborn wanted to fill. Along came Roy Chapman Andrews, a relatively new employee of the American Museum, who fed into the belief system of Osborn and proposed a venture to the Gobi Desert of Mongolia. Andrews, a biologist by training, organized the largest American expedition to ever leave the States. His field crew was unique because he brought along many different scientific disciplines to supplement one another. He was also one of the first to use Dodge cars and Fulton trucks for the expedition staff. His planning and logistics were meticulous, sending camel caravans ahead to provide the necessary gasoline and supplies. The camels also carried back the fossils that were discovered.

In the Flaming Cliffs, 300 miles south of Urga, Mongolia, one of his scientists discovered a clutch of dinosaur eggs, which they believed were laid by the new species they also discovered nearby, *Protoceratops*. After finding embryos in similar eggs, years later, scientists now know the eggs were laid by the theropod, *Oviraptor*.

Theropods (presumed meat-eaters) like *Velociraptor* and *Oviraptor* were also discovered in Mongolia by the Andrews' expeditions. In the *Jurassic Park* movie series, a larger dinosaur (*Deinonychus*) was used to look more frightening, but was still called *Velociraptor*, which was actually only the size of a large turkey.

Several *Oviraptor* specimens have been found on or near its own eggs. This dinosaur is classified as a theropod, but had no teeth. It was about six feet long. It was originally thought to have eaten the eggs of *Protoceratops*, so its name means "egg stealer."

Some claim that Roy Chapman Andrews (1884–1960) was the model for the movie character Indiana Jones. He wore a similar broad-brimmed hat, often sported a rifle, and had a zest for life and the dramatic. He made the cover of *Time* for his discoveries in 1923.

Andrews hoped to find the human "missing link" for his boss Osborn, but instead his team found a "gold mine" of dinosaurs, and most remarkably, the first recognized dinosaur eggs. Today, evolutionists are still coming up empty in their search for the elusive missing links to fill in the gaps in their faulty theory.

The Flaming Cliffs site, most famous for yielding the first discovery of dinosaur eggs. Other finds in this area include specimens of *Velociraptor*, *Oviraptor*, and *Protoceratops*.

Dinosaur egg fossil excavation. Danish-born paleontologist George Olsen showing his find of fossil dinosaur eggs to expedition leader, U.S. explorer, and naturalist Roy Chapman Andrews (1884–1960). Photographed during the 1925 expedition to Mongolia by the American Museum of Natural History.

Roy Chapman Andrews, who many believe was the inspiration for "Indiana Jones," was most famous for the discovery of the first fossilized dinosaur eggs. The eggs were originally thought to be from *Protoceratops* but were later determined to be from *Oviraptor*.

DINOSAUR RENAISSANCE

Dinosaur discoveries slowed down during the late 1930s through the 1950s due to World War II and the tension of the Cold War. However, in the late 1960s, Dr. John Ostrom of Yale University was back out in Montana with his paleontology students. One of these undergraduate students happened to be the future Dr. Robert T. Bakker. Dr. Ostrom unearthed a new species near the Wyoming and Montana border that he described and named in 1969 as *Deinonychus*. The name means "terrible claw" because of the large claw found on the second toe of the hind foot. The stiffened, hollow tailbones placed it in the same subgroup of theropods as *Oviraptor* and *Velociraptor*. This dinosaur was portrayed as "*Velociraptor*" in the *Jurassic Park* movies because the name sounded more intriguing to the Hollywood movie producers.

In his 1969 paper describing *Deinonychus*, Dr. Ostrom argued that it was active and agile, and had the ability to run and move quickly. This may not be surprising to most of us today, but it was shocking to most paleontologists in the early 1970s. Since the time of Marsh and Cope, dinosaurs were portrayed as sluggish and slow reptiles that "shuffled" around swamps. Paintings by dinosaur artists, like Charles R. Knight, presented this lumbering and bumbling image to the public. Ostrom's radical ideas set off new ways of thinking about dinosaurs for a whole new generation of paleontologists. It was the start of a dinosaur renaissance in science. Soon thereafter, Dr. Robert T. Bakker and others were insisting that dinosaurs, like *Brontosaurus*, were warm-blooded. More new discoveries in the 1970s and 1980s seemed to even hint at aspects of dinosaur behavior.

Creationists agree that dinosaurs were fast and agile creatures. Whether or not they were warm or cold-blooded is still unclear. They must have been fearful animals that most pre-Flood humans likely stayed away from. God described the Behemoth to Job as "first among the works of God" (Job 40:19; NIV).

Dr. John Ostrom was also an expert on flying reptiles. On one of his trips to Europe he "discovered" an *Archaeopteryx* specimen misidentified as a pterosaur by examination of the upper arm length.

Two of America's most prominent dinosaur paleontologists, Dr. Robert Bakker and John (Jack) Horner, have both made important discoveries and contributed to the dinosaur renaissance. Horner discovered the first eggs in the USA in Montana in 1978, and named a new species in the process, *Maiasaura*, the "mothering dinosaur." Horner found evidence suggesting these dinosaurs may have stayed near their hatchlings for a few days or weeks, tending to their needs. Although Horner never finished his college degree, he has advised many Ph.D. candidates, including Dr. Mary Schweitzer. Horner was also the technical adviser to the *Jurassic Park* movie series.

Thermopolis specimen of *Archaeopteryx lithographica* from the Solnhofen Limestone of southern Germany. The Haarlem specimen of *Archaeopteryx* was originally misidentified as a pterosaur.

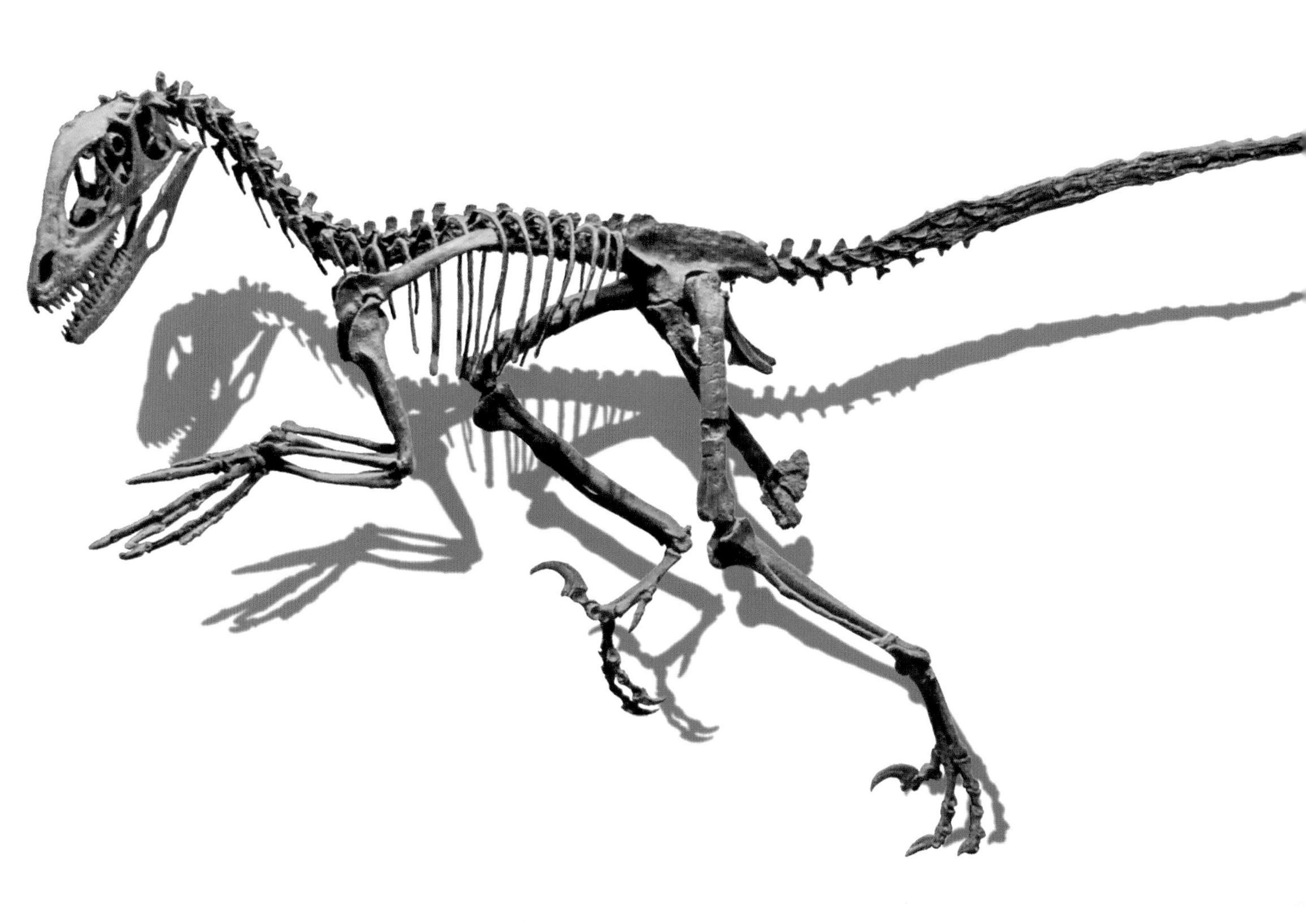

Deinonychus claw (top) and *Velociraptor* claw (bottom).

Deinonychus, like many "raptors," had a large claw on its hind foot. However, recent experimental studies have shown that the claw was not capable of slashing as portrayed in the movies. It doesn't possess a sharp enough underside to make long cuts. It was, however, very effective at puncturing. These large hind claws may have been used to help grab onto prey or to kick and puncture the neck arteries and/or windpipe.

Dinosaurs are now viewed as fast-moving, agile reptiles that thrived in the pre-Flood world. How long they survived after the Flood and why they went extinct will be discussed in a later section in this book. Whether or not they were warm or cold-blooded like reptiles today will also be addressed.

Theropod Dinosaur (*Allosaurus fragilis*). Jurassic.
Dinosaur National Monument, Utah, USA.

CHAPTER 5:

THE MANY VARIETIES OF THEROPOD DESIGN

INTRODUCTION TO DINOSAUR DESIGN

Recall from chapter 1 that dinosaurs can be classified into two orders, based on hip style: the Saurischia and the Ornithischia. These orders are divided into five suborders, and then further classified into 13 infraorders and 60 families. Many creation scientists think a family may approximate a "kind." In the next five chapters, we will review each of the five suborders, many of their associated infraorders, and even a few genera.

We will start our review of the dinosaur types with the two Saurischian dinosaur orders, or the lizard-hipped dinosaurs. These two suborders, the Theropods and the Sauropodomorphs, make strange "bedfellows," but are lumped together, as they have similar hip design. After that, we will discuss the remaining three suborders within the Ornithischian order: the Ornithopods, the Marginocephalia, and the Thyreophorans. And as before, we will look at example genera within each category. Throughout, you will notice great diversity within the various kinds. God built in plenty of variety in the dinosaurs.

SUBORDER THEROPODA: THE MEAT-EATERS AND MORE

The name Theropod means "beast-foot." God created this group of dinosaurs unique in several ways. First, Theropods were designed with larger eyes and bigger brains than most dinosaurs. Most walked on two feet, had extra holes (fenestra) in the skull to reduce weight, and possessed a jointed jaw and skull for flexibility in biting and chewing. Many of the theropod bones were hollow, and the head was attached to the neck by a very mobile joint. Oddly, the theropods possessed a lizard-hipped pelvis, even though many secular paleontologists think they are closely related to birds. Theropods are also the rarest of all dinosaur groups, making up less than 5 percent of dinosaur discoveries.

This is also one of the most difficult groups to classify, as there is such inherent created diversity in the fossils. Evolutionary scientists are constantly reshuffling the theropod groupings, trying to fit them into some forced evolutionary pattern. The classifications are therefore tentative and subject to constant change. They fail to notice that the only consistent pattern they adhere to is not based on actual specimens, but on a preconceived and distorted evolutionary mindset. The rocks show that each theropod dinosaur appears fully formed, without ancestors, and then quickly disappears, just as you would expect in successive deposits of a global flood. There is not one single discovered and/or undisputed transitional fossil, or intermediate, between the theropod groups.

Most of the theropod dinosaurs fall into two groups: 1) Ceratosaurs and 2) Tetanurae. The Ceratosaurs were created with many fused leg bones that allowed for fast running. They commonly had horns or a crest along the top of the skull, giving meaning to their name, the "horned lizards." Included in this group are the dinosaur genera *Coelophysis* and *Dilophosaurus*, and possibly *Herrerasauria* also.

The Tetanurae ("fused-tail lizards") were a more diverse group and includes two infraorders: 1) Carnosaurs and 2) Coelurosaurs. Tetanurae were given a second opening in the skull to further lighten the skull weight and a differently designed hind leg to allow even faster running and quicker movement. One family of Carnosaurs ("meat lizards") is recognized (Allosauridae) and in some classifications includes *Allosaurus*, *Megalosaurus*, and possibly *Spinosaurus*. The Coelurosaurs are divided into at least three families, the Tyrannosaurs, the Ornithomimosaurs, and the Dromaeosaurs. The Dromaeosaurs include the raptors like *Velociraptor* and *Deinonychus*. Some also lump the Therizinosauridae into

Giganotosaurus skull

this group, but the incomplete nature of most of these fossils makes classification difficult.

Giganotosaurus, found in Argentina in the mid-1990s, is one of the largest meat-eating dinosaurs discovered thus far. It was about 43 feet long from the tip of the snout to the tip of the tail. The teeth were designed smaller and thinner than *T. rex* teeth. Many secular paleontologists think its design fits in the Allosauridae group. But its overall body shape and design make it look very much like other large predatory dinosaurs. It is possibly the same biblical kind as the *T. rex* and the *Allosaurus,* in spite of evolutionary-driven classifications.

Dinosauria	Saurischia	Theropoda	Herrerasauria
			Ceratosauria
			Carnosauria
			Coelurosauria
			Therizinosauria
		Sauropodomorpha	Sauropoda
			Prosauropoda
	Ornithischia	Ornithopoda	Fabrosauria
			Ornithopoda
		Marginocephalia	Pachycephalosauria
			Ceratopsia
		Thyreophora	Stegosauria
			Ankylosauria

WAS LEVIATHAN A *SPINOSAURUS*?

Specimens of *Spinosaurus* have been found mostly in northern Africa. It is unclear if they should be classified as true Carnosaurs or if they should be placed in another theropod group. This dinosaur was made famous by the *Jurassic Park* movies as the dinosaur that defeated *T. rex*. Whether an encounter between the two would have had the same result in reality is highly unlikely. I personally think the strong bite of *T. rex* would have prevailed, but that's just my opinion. The latest estimate has this dinosaur a bit longer than *T. rex*, coming in at 50 feet long from snout to tail.[1]

Spinosaurs were designed with a large bony sail on their backs up to seven feet high. The exact purpose of this sail is unclear. Recent research indicates that spinosaurs spent a considerable part of their lives in water, possibly eating fish and other aquatic prey. They had long, narrow jaws with round, reptile-like teeth in the lower jaw, and larger, more dinosaur-like teeth in their upper jaw. They seem perfectly designed to live in aquatic settings like crocodiles today, possibly possessing paddle-like hind feet and nostrils placed high on their skulls.[2]

The Leviathan described in Job 41 may be a description of a semiaquatic dinosaur like the *Spinosaurus*.

> Who can open the doors of his face, with his terrible teeth all around? —Job 41:14
>
> Darts are regarded as straw; He laughs at the threat of javelins. His undersides are like sharp potsherds; He spreads pointed marks in the mire —Job 41:29–30
>
> He leaves a shining wake behind him; One would think the deep had white hair. On earth there is nothing like him, which is made without fear —Job 41:32–33

These descriptive words indicate a very large, meat-eating reptile that at least could crawl or walk on the bottom of a river or shallow water body, similar to the suggested behavior of *Spinosaurus*.[3] The "wake behind him" phrase may be referring to the sail that would have stuck up out of the water along its back. It seems possible that God was making His point using two of the largest dinosaurs known to Job in chapters 40 and 41. One was a large plant-eater (behemoth), and the other was the "king over all the children of pride" (Job 41:34). Both of these animals could have been right there in or near rivers where Job resided.

Nizar Ibrahim et al. reported that sharks, sawfish, ray-finned fishes, and coelocanths were found in the same rock layers as a *Spinosaurus* in Morocco. These ocean animals found mixed with land animals are difficult for secular scientists to explain. Today's

Spinosaurus **skeleton showing jaw and spine**

coelocanths live about 500 feet below the ocean surface and not in freshwater rivers as paleontologists have proposed. Secular scientists dismiss the blatant physiological evidence from living specimens and insist that ancient coelocanths must have lived in fresh water simply because they are found in strata with dinosaurs. Where is the logic in this conclusion?

Dinosaur fossils found in rock strata mixed with marine fossils are commonplace and not the exception. Empirical evidence cannot be ignored or simply explained away as a rare occurrence. The fossil evidence supports a catastrophic and global flood that mixed the marine realm with the terrestrial realm as tsunami-like waves spread ocean fauna and sediments across the continents. Genesis 7 and 8 describe this process better than any secular scientist could imagine.

INFRAORDER CERATOSAURS: DID *DILOPHOSAURUS* REALLY SPIT?

Dilophosaurus means "two-crested reptile" and was first described in 1954 from partial skeletons found in the Lower Jurassic system Kayenta Formation in Arizona in 1942. Sam Welles originally named this dinosaur *Megalosaurus,* thinking it was similar to the previously-named British dinosaur. However, in 1970, after finding more skeletal remains at the site in 1964, he renamed the dinosaur *Dilophosaurus*. This species is distinguished by the dual, thin, bony crests that run along the roof of the skull. The adults reached about 20 feet in length and are estimated to have weighed between 1,000 and 2,000 pounds. Its sharp, blade-like teeth show a distinct gap in the upper jaw between the front (premaxillary) teeth and the cheek teeth (maxillary). *Dilophosaurus* is classified in the infraorder Ceratosauridae ("horned lizards") due to its crested skull. Although shown in the movies, there is no scientific evidence to place a retractable neck-frill around the neck of *Dilophosaurus* or any dinosaur.

Although *Dilophosaurus* was a biped, it had rather long upper arms. The hand had four fingers, but only three of the fingers had a claw attached. The dual crests that ran the length of the skull were quite thin, and they were probably designed for ornamentation or sexual display and not for any sort of protection.

Dilophosaurus is shown spitting some sort of poison on its prey before devouring it in the movie Jurassic Park. But how realistic is this scenario? Is there any scientific basis for this behavior in dinosaurs? Although many dinosaur soft tissues have been found, such as proteins, collagen, bone cells, and red blood cells, no soft tissue evidence of a "spitting" dinosaur has yet been identified. However, there are multiple examples today of reptiles and mammals that spit or spray some sort of toxic venom or foul-smelling liquid.

Theropod footprint at the Purgatoire River dinosaur track site, Picket Wire Canyonlands, Colorado, U.S.A.

World's smallest dinosaur tracks found at Wasson Bluff, Nova Scotia, Canada.

In Job 41:19–21, God describes the leviathan as being able to spit "sparks of fire." So whether or not *Dilophosaurus* could or could not spit, it is possible that some dinosaurs did spit venom or even some type of "fire."

The Kayenta Formation in Arizona contains much more than just the dinosaur remains of *Dilophosaurus*. Paleontologists have found numerous other fossils mixed in the same rocks with the dinosaurs. Oddly, many of these other fossils are water related, such as sharks, bony fish, bivalves, and snails. In addition, they have found crocodiles, turtles, mammals, other dinosaurs—and even a pterosaur (flying reptile). Evolutionists have a tough time trying to explain these mixed ocean and land assemblages. How can land animals coexist in a marine setting and vice versa? Creation scientists can easily explain this apparent uniformitarian dilemma. During the great Flood, the Kayenta Formation was formed by tsunami-like waves mixing various animals from the ocean and land with sands and silts and spreading them in a deposit across what is now Arizona. This resulted in the catastrophic and well-mixed siltstone layer we see today.

Footprints attributed to *Dilophosaurus* have been found in rocks as far away from Arizona as Sweden. However, most of these have now eroded away since the 1970s. Preservation of trackways and dinosaur footprints must take place quickly, immediately after they are made by the living animal, or they too quickly disappear. The multitude of dinosaur footprints found all over the world in many rock layers is concrete evidence of a rapid burial. Creation scientists conclude the burial of these trackways occurred during the year-long, global Flood.

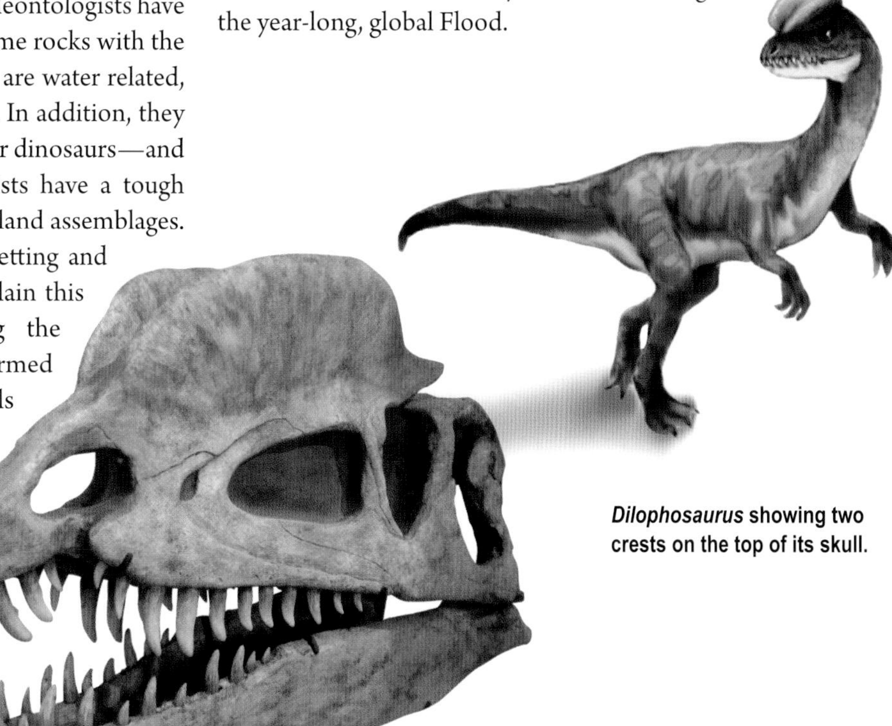

Dilophosaurus **showing two crests on the top of its skull.**

Dinosaur track in Tuba City

Large theropod footprint. Jurassic. Loulle (Jura, Franche-Comté, France)

INFRAORDER CARNOSAURS: *ALLOSAURUS*: THE LAND SHARK

Although *Allosaurus* didn't get a lot of screen time in the *Jurassic Park* series, it was one of the more common theropod dinosaurs throughout the American West. Numerous partial and near-complete specimens of this dinosaur have been found in the Morrison Formation in locations across Wyoming, Utah, Colorado, Montana, and New Mexico. *Allosaurus* means "different reptile," but this dinosaur was not always known by this name. In 1870, Dr. Joseph Leidy named a partial dinosaur skeleton *Antrodemus,* meaning "body cavity." A different specimen of this same dinosaur found near Cañon City, Colorado, was later called *Allosaurus* by O.C. Marsh in 1877. It wasn't until the mid-20th century that the names were combined. Most scientists place *Allosaurus* in the clade Tetanurae ("fused tails") and in the infraorder Carnosaur ("meat reptile").

The adult *Allosaurus* reached a length of about 30 feet, but some paleontologists have found partial specimens that may have approached 40 feet in length. The weight of *Allosaurus* is estimated anywhere from 1.5 to 2.5 tons. Compared to *Tyrannosaurus rex,* the *Allosaurus* had rather small teeth for its size. The teeth only stuck up about 2 inches above the jaw line, but were flatter than *T. rex* teeth and very sharp, similar in shape to the tip of a sabre. Because of this, it is assumed that this dinosaur hunted like sharks. *Allosaurus* could use its teeth to slash at prey and cause horrific loss of blood, weakening the victim. Then the *Allosaurus* could move in for the kill. Computer models have suggested that the bite of an *Allosaurus* was much weaker than the *T. rex* and even much weaker than a modern alligator. Other computer simulations indicate that *Allosaurus* could run about 21 mph, a bit faster than the estimates for a *T. rex*. These data seem to support that *Allosaurus* was a slasher, using its head like a hatchet to cause bleeding and weaken its prey.

There is little strong evidence to suggest that *Allosaurus* was a pack hunter, other than finding multiple specimens buried together by transport, like at the Cleveland-Lloyd Quarry. Likewise, there is also little fossil evidence to indicate *Allosaurus* was a solitary hunter. Some may have hunted in small packs and others alone. Finding multiple individuals at a site merely means they were in close proximity at the time of burial. The rising Flood waters may have herded them together prior to burial, but that doesn't mean they necessarily lived together in more "normal" times.

Many "shed" *Allosaurus* teeth can be found with and near sauropod dinosaurs. I found two such teeth near the tail vertebrae of an *Apatosaurus* in Wyoming. Does this mean the *Allosaurus* was eating the sauropod? Not necessarily, as the teeth and bones were found in a bone bed with hundreds of other bones of various dinosaurs, including a partial specimen of an adult *Allosaurus*. These dinosaurs were transported within a large sediment flow as part of the great Flood and deposited

An illustration of *Allosaurus*

Allosaurus fragilis over *Apatosaurus*, American Museum of Natural History

"log jam" style in a jumbled mass of bone and debris, including many logs and plants.

Allosaurus did seem to hunt live prey before the Flood, and in particular, stegosaurs. Paleontologists have found examples of injuries to an *Allosaurus* that matched the tail spike of a *Stegosaurus*, and of bite marks on a *Stegosaurus* that fit the teeth pattern of an *Allosaurus*. Because CT scans of the brain cavity indicate a close resemblance to the modern alligator brain in shape and dimensions, *Allosaurus* likely attacked like an alligator, snapping at anything that it perceived as food.

One of the most complete *Allosaurus* specimens was discovered in the Bighorn Basin of Wyoming in 1991. It is known as "Big Al" and is on display at the University of Wyoming Geological Museum. Paleontologists estimate it was six years old, 26 feet long, and weighed about 3,300 pounds. It had sustained numerous injuries, including cracked ribs, vertebrae, and a splintered hand bone that had healed, and also had a significant bone infection in the right hind foot at the time of its death and burial in the Flood. These animals seem to have participated in the pre-Flood violence we read about in the Book of Genesis.

At least 44 adult specimens of *Allosaurus* were found buried together in one spot in central Utah. Known as the Cleveland-Lloyd Quarry, this site has been under excavation since the 1920s, but it wasn't until the 1960s that many thousands of bones of *Allosaurus* were extracted. Secular paleontologists still struggle trying to explain this unusual concentration of disarticulated and well-mixed bones of all sizes and ages. In contrast, creation scientists explain this as a massive sediment flow that mixed and tore apart the bodies and bones of the 44 allosaurids and dumped them in one condensed deposit during the great Flood.

AIG'S *ALLOSAURUS* "EBENEZER"

In 2014, Answers in Genesis unveiled a full-grown allosaur at the Creation Museum in Kentucky, dubbed Ebenezer, after the "stone of help" in 1 Samuel 7. The specimen was discovered by Dana Forbes in 2000 within beds of the Morrison Formation (Jurassic system) in northwestern Colorado.[2] Ebenezer is the largest dinosaur on display at the AIG museum at 10 feet tall and 30 feet long. It was estimated to have weighed about 1.8 metric tons at the time of death in the Flood.[4]

Although only about half the bones of Ebenezer were found at the site, most of the skull and neck were intact and still articulated. In fact, it is one of the most complete skulls ever found of *Allosaurus*. All 53 teeth were still in the jaws, averaging 4.5 inches (11.4 cm) long, including the roots. Ebenezer's head and neck were found broken off from the rest of its body as a testament to the great forces involved in transport and/or burial.

Oddly, Ebenezer was found alone and not with other allosaurs, or even another dinosaur, as is so common elsewhere in the Morrison Formation. AIG scientists interpret the sandstone matrix surrounding the specimen as part of a massive flow that occurred during the great Flood, entombing and preserving Ebenezer until its recent discovery.[5] The reconstructed specimen is a reminder of the catastrophic judgment of the sinful and violent pre-Flood world.

Photos courtesy of Answers in Genesis.

T. rex Sue skeleton

INFRAORDER COELUROSAURS: *TYRANNOSAURUS REX* "SUE"

The most famous, largest (at 41 feet), and most complete (over 90 percent was recovered) *Tyrannosaurus rex* ever found was spotted by a paleontologist named Sue Hendrickson in the summer of 1990. On August 12, Sue was walking around lost in fog in western South Dakota on a crew working for the Black Hills Institute of Geological Research out of Hill City, South Dakota. She came back to camp with two vertebrae bone fragments showing a distinctive honeycomb texture called camellate structure. These hollowed bone features were indicative of a theropod dinosaur. The only question was, what type? The dinosaur she found was big, estimated to have weighed between 5 and 7 metric tons when alive! It turned out to be the most spectacular and publicized *T. rex* discovery of the 20th century.

The *T. rex* specimen was soon nicknamed "Sue" to honor its discoverer, as was the tradition of the Black Hills Institute. As the scientists excavated the dinosaur, they found it fairly well articulated and nearly complete. They also found bones from two smaller, juvenile *T. rex* specimens along with the bones of Sue. Because it was found on Indian Reservation land, the federal government stepped in and confiscated the entire skeleton and locked it away for three years while a court battle for ownership ensued. Eventually, the case was resolved and the specimen auctioned off to the highest bidder by Sotheby's. A scramble went through the U.S. scientific community to gather funds to make a bid. There was worry that a foreign museum would buy the skeleton and that U.S. scientists would lose access to the most spectacular *T. rex* ever pulled from U.S. "soil." The Field Museum of Natural History in Chicago, in collaboration with McDonald's and Disney placed the winning bid of $8.36 million, of which $7.6 million went to the owner of the reservation land, and Sotheby's got the rest. Sue Hendrickson received nothing for her discovery and excavation efforts except more than $100,000 in legal bills.

Possibly the most complete skull of a *T. rex*, even better preserved than Sue's skull, was found by another member of the Black Hills Institute, Stan Sacrison. This specimen gave scientists additional details of the brain cavity of *T.rex*, demonstrating its large olfactory lobe (for smell) and its overall shape that was similar to modern alligators and not like birds.

Most theropod dinosaurs, including Sue, had a "floating" frontal rib cage called a gastralia that protected the abdomen region from injury. These bones were not attached to the vertebrae like normal ribs. Unfortunately, these bones are rarely shown in museum reconstructions, as they need a lot of struts and tend to take away from the display appearance. *T. rex* Sue's gastralia is off to the side of the reconstructed skeleton in a separate showcase at the museum.

Sue had foot-long teeth and large olfactory bulbs, but was the *T. rex* a predator, scavenger, or opportunist? Most paleontologists think of *T. rex* as an opportunist, eating anything it could obtain, dead or alive. A few scientists, like John R. Horner, think *T. rex* was exclusively a scavenger due to its great sense of smell. However, healed bite marks, attributed to an animal the size of a *T. rex*, had been found on both a *Triceratops* and an *Edmontosaurus* skeleton, indicating they survived a predatory attack. But were these bite marks really from a *T. rex*? The matter was finally resolved in 2013 when scientists reported finding the tip of a *T. rex* tooth embedded in the backbone of a duck-billed dinosaur. The backbone had healed around the tooth fragment, demonstrating the hadrosaur had survived for some time after the failed attack. This supports the notion that *T. rex* was a predator at least some of the time.

In God's original creation, even the mighty *T. rex* was a vegetarian, as were all animals. It wasn't until after the sin of man and the resulting Curse that *T. rex* became a meat-eater. Exactly what type of vegetation *T. rex* ate with those massive teeth remains a mystery, but animals with similarly sharp teeth have remained exclusively vegetarians, like the fruit bat and panda. A recent study of modern crocodilians has shown that nearly three-fourths of them consume plants, including fruits, nuts, and grains to supplement their diet, leaving secular science baffled for an explanation.[6]

The 2015 discovery of a theropod dinosaur named *Chilesaurus* has created another stir in secular science. This dinosaur was found in southern Chile in Flood strata from the Upper Jurassic system. It was classified as a Tetanurae theropod based on its overall skeletal structure, but had teeth like a plant-eating dinosaur. The scientists concluded that plant-eating among theropods was more commonplace than previously thought.[7]

Plant consumption by predators continues to surprise many evolutionary scientists, but it does not surprise creation scientists. *Tyrannosaurus rex* may have had the strongest bite and the biggest teeth, but it could, and did, survive in the past on a diet of plants alone. The Bible is confirmed again by the findings of science.

Close up of the skull of *Tyrannosaurus rex* "Sue" at the Field Museum of Natural History in Chicago. This is a replica skull on the body. The actual skull is in a separate showcase on the next level. Note that both eyes can look straight forward, indicating that *T. rex* likely had stereoscopic vision.

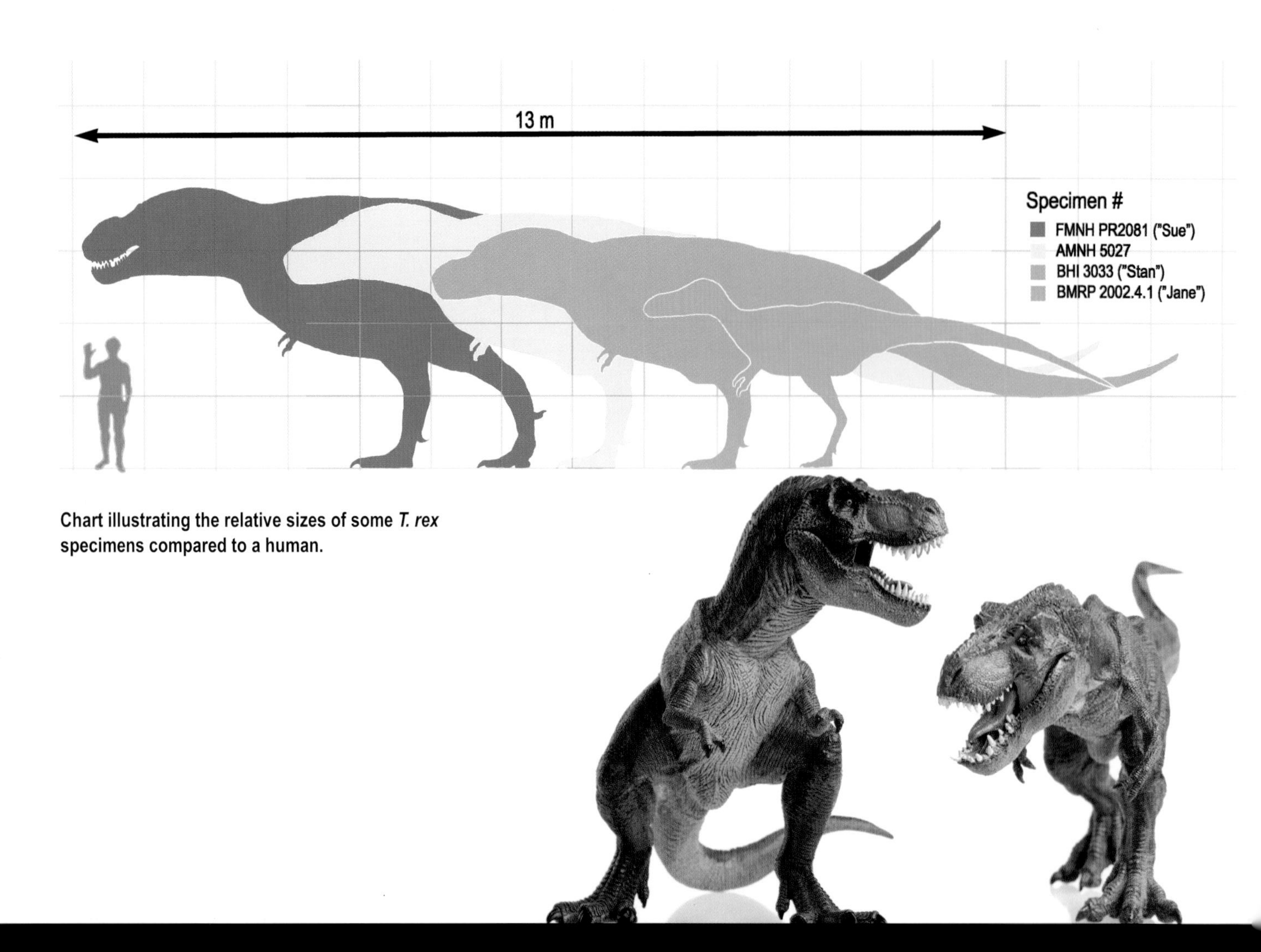

Chart illustrating the relative sizes of some *T. rex* specimens compared to a human.

WAS *T. REX* SUE A MALE OR FEMALE?

Peter Larson determined that Sue fit the larger, robust morphotype, making it bigger than the gracile, more slender variety. But which variety was the male and which the female? Were the males bigger or the females bigger? His initial examination of Sue showed a missing chevron (a bone that sticks downward from the tail vertebrae) between the first and second caudal vertebrae, mimicking the bone structure of many female crocodiles. This seemed to suggest Sue was a female, and that larger *T. rex* specimens were females and smaller, gracile varieties, were male. However, after fully excavating the skeleton, a chevron was found between the first and second caudal vertebrae, implying Sue was instead a large male. Unfortunately, the crocodile analogy doesn't always hold true, and without the presence of medullary bone in the legs, there is no conclusive method at present of determining if Sue was male or female. The lesson here is — don't judge a dinosaur by its name. In chapter 10, I will address some evidence supporting that Sue was likely a male dinosaur.

How old was Sue? The skeleton exhibited many healed injuries to the ribs and right side of the body, including the right humerus and pectoral girdle region (upper arm). Yet scientists found no conclusive evidence of bites by another *T. rex* to the skull or body. Based on the number of injuries and healed bones, it seems likely that Sue was very old. One study counted growth rings in the cheek and rib bones of Sue and calculated an age of 28 years old. However, growth rings in dinosaurs have been shown to be unreliable and may only give a minimum age for dinosaurs. So it seems likely that Sue was much older than 28, and may have been well over 100.

INFRAORDER COELUROSAURS: *VELOCIRAPTOR*: THE LITTLE THREAT

Few dinosaurs strike fear in adults and children like *Velociraptor* does in the *Jurassic Park* series of movies. Unfortunately, they were portrayed by a "body-double" in the movies using a dinosaur found on a completely different continent. *Deinonychus*, a theropod dinosaur found in North America, just didn't have a "catchy" name, so the producers scratched it from the movie credits. *Velociraptor* was substituted instead, but in name only. *Velociraptor* was really much smaller, much less intelligent, and probably nowhere near as vicious. Many *Velociraptor* specimens have been found in Cretaceous system strata in central Asia, beginning with their discovery by the Central Asiatic Expeditions.

Velociraptor was about 6 feet long from nose to tail, but its body was only about the size of an extra-large turkey, weighing about 35 pounds. *Deinonychus*, by contrast, was about 10 or 11 feet long and weighed about 150 pounds. However, biomechanical analysis has *Velociraptor* running at top speeds of around 24 mph, faster than most humans. Maybe it's a good thing humans stayed far away from these dinosaurs in the pre-Flood world.

Both of these dinosaurs belong to the Dromaeosaur family. Although the brain-to-body ratio of the dromaeosaurs was one of the largest of all dinosaurs, it was still less than an equally sized mammal like a dog. *Velociraptor* probably could not open doors with purpose and "learn" new behavior easily. Its brain was like all theropod dinosaurs, closer in shape to that of a modern alligator, with a small brain section for processing information and a large section for sensory perception. It would likely strike at anything it thought was a food source when hungry. Pack-style hunting seems likely, but probably far less coordinated than is implied by the movies.

All adult dromaeosaurs, including *Velociraptor*, were designed with a stiffened and straight ossified tail that was flexible only just behind the pelvis. Muscle attachment sites indicate the tail swung from side to side with each step, balancing the bipedal gait.

Velociraptor was designed with an enlarged claw on the second digit of its hind foot. This claw is portrayed in the movies as having the ability to slash and disembowel prey. However, studies have demonstrated the claw merely punctures and was not sharp enough on the lower side to slash through tough skin. The two enlarged claws on the hind feet may have been used to grasp or hold on to its prey, especially if it jumped on a larger animal.

Velociraptor had a mouthful of small teeth that were sharp and serrated and curved backward, preventing prey from escaping. These presumably could slash at prey and cause significant bleeding, weakening the prey in shark-style.

Examples of *Velociraptor* illustrations

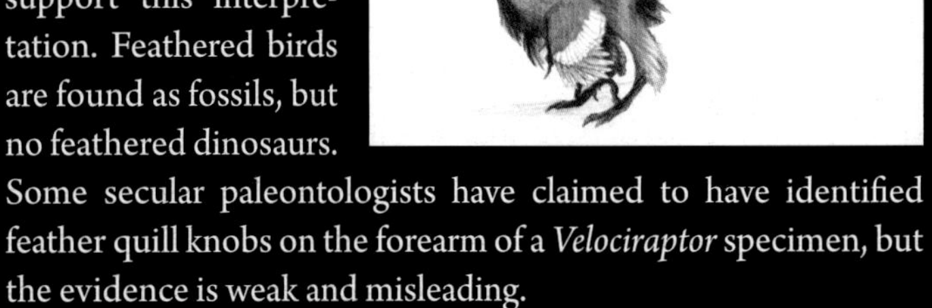

Modern reconstructions of *Velociraptor* often include feathers on part of its body. There is absolutely no compelling evidence to support this interpretation. Feathered birds are found as fossils, but no feathered dinosaurs. Some secular paleontologists have claimed to have identified feather quill knobs on the forearm of a *Velociraptor* specimen, but the evidence is weak and misleading.

A three-inch bone shard from a pterosaur (a flying reptile) was found fossilized within the ribcage of one specimen of *Velociraptor*. The dinosaur probably died shortly after swallowing the pterosaur bone, as the outside of it showed no erosion by stomach acid. This little theropod likely ate a "last meal" and was soon trapped by the rising Flood waters and rapidly buried in sediment, preserving the evidence inside.

INFRAORDER COELUROSAURS: HERBIVORE THEROPODS?

Although most theropods had sharp, serrated teeth, some were designed with no teeth at all, like many oviraptorosaurs and ornithomimosaurs. Even the therizinosaurids had leaf-shaped teeth similar to many plant-eating dinosaurs.

In the past, oviraptorosaurs were thought of as the "egg-thief" dinosaurs. They gained this reputation because their skeletons were found in the 1920s in Mongolia near what was once thought to be *Protoceratops* eggs. The formal name of *Oviraptor philoceratops* translates as "egg thief that loves ceratopsians."

In the 1990s, paleontologists discovered that *Oviraptor* was not the egg-eating animal it was portrayed to be. An embryotic *Oviraptor* was found in one of the eggs near an adult *Oviraptor*, suggesting it was the parent of the eggs. Another spectacular specimen showed an adult *Oviraptor* buried and fossilized right on top of an egg nest. Evolutionary scientists claim that this "mother" dinosaur was incubating the eggs, but it is more likely that the dinosaur was merely caught in the act of laying eggs and became trapped atop the nest as Flood waters and sediment engulfed them.

Oviraptor is now thought to have eaten plants or possibly clams with its strange, toothless mouth and "beak." After careful examination of the jaw structure, other scientists question whether *Oviraptor* had strong enough jaws to crack open shellfish. Because one *Oviraptor* was found with a small lizard fossil in the stomach region, many paleontologists assume these dinosaurs were omnivores or herbivores that may have supplemented their diet with an occasional meaty snack.

Oviraptor was discovered by Roy Chapman Andrews on one of the early Central Asiatic Expeditions to Mongolia in the 1920s. It was described and named by Henry F. Osborn in 1924. An adult was about 6 to 8 feet long and only weighed about 55 to 75 pounds. Found in Upper Cretaceous system rocks (late Flood), it is thought to be a swift runner, as it had long hind legs and long shin bones. Some have estimated its top speed at 40 mph.

INFRAORDER COELUROSAURS: ORNITHOMIMOSAURIDS ARE HERBIVORES TOO?

Initially called *Ornithomimus* by O.C. Marsh in 1892, *Struthiomimus* "ostrich mimic" found its lasting name in 1917 after Henry F. Osborn acquired new specimens from Cretaceous system (late Flood) rocks in Alberta. This long-legged dinosaur was about 13 feet long and weighed an estimated 310 pounds. It had a long, slender neck and a small skull, with large eyes and a toothless beak covered with keratin. These dinosaurs were also designed with long, slender forelimbs with large claws.

Gallimimus was another ornithomimosaurid dinosaur found buried in Flood sediments in Mongolia. Similar to *Struthiomimus,* and likely the same kind of dinosaur, they are thought by many paleontologists to be the fastest running dinosaurs. God designed them with long metatarsal bones tightly bound together in the foot. In addition, they had long shinbones, allowing for top speeds of close to 50 mph.

Many paleontologists consider Ornithomimosaurids to have been herbivores because most had no teeth and are occasionally found with gastroliths in their stomach region. The gastroliths were presumably for assistance in grinding up swallowed vegetation.

INFRAORDER THERIZINOSAURIDS: DON'T ASK!

Therizinosaurids have been described as a dinosaur "designed by committee." Most paleontologists now place them in the theropod suborder, but they have been placed with other groups in the past, including an older classification as Segnosaurs. *Therizinosaurus* was named by Russian paleontologist Evgeny A. Maleev in 1954 after discovering the specimen in Cretaceous system rocks in Mongolia in 1948. He named it "scythe reptile" after the enormous 3-foot claws found on the forelimbs. Some scientists speculate it may have used these claws to tear open termite or ant nests, suggesting it was an insectivore. A more recent study of the claws found they were more likely suited for digging roots and/or grasping branches.[8] Most paleontologists assume they were herbivores or omnivores based on the shape of their teeth. These dinosaurs grew up to 23 feet long and 10 feet tall, and are estimated to have weighed 3 tons.

Articulated *Plateosaurus* skeleton, a prosauropod.

CHAPTER 6:

The Sauropodomorphs were the longest and heaviest animals to ever walk the earth. Their leg bones were designed thick and strong. Some were at least 140 feet long and possibly 160 feet. Many are commonly thought to have weighed as much as 40 or 50 metric tons, with some estimates as high as 100 metric tons. Features that made them unique were the presence of ten or more elongated vertebrae in the neck, giving them the title of "long-necks," and most had an equally long tail. Their gigantic weights were supported by massive leg bones and a thickening of the cartilage between the joints for extra cushioning. These dinosaurs had rather small heads for their body size. They also had short, compact feet, like elephants, with a large claw on the first digit of the front feet (the big toe). All of these dinosaurs are thought to be herbivores, as few had sharp teeth and some only had teeth in the front of their mouths. Sauropodomorphs are divided into two infraorders: 1) Prosauropods and 2) Sauropods.

Plateosaurus

Prosauropods were designed smaller and shorter than most Sauropods. Many were less than 25 feet long. Some may have even walked bipedally. Prosauropods also differed from Sauropods because the fifth digit on the hind foot was very small. Evolutionists claim this small toe was "rudimentary," but there is no evidence to back this assertion. Prosauropods were fairly abundant in the pre-Flood world, as their fossils have been found on six continents, including Antarctica.

One of the most common varieties of Prosauropod is the genus *Plateosaurus*, found throughout Western Europe. This animal may not have been an exclusive herbivore, as the teeth were leaf-shaped and even serrated. It may have been omnivorous.

Sauropods are the more famous of the Sauropodomorphs. They include most of the famous long-necks, such as *Brontosaurus* or *Apatosaurus*, *Brachiosaurus*, *Diplodocus*, and *Camarasaurus*. Sauropods were designed with holes on the side of their back vertebrae, called pleurocoels. These openings lessened the weight of the vertebrae, like the fenestra in the skulls of theropods, lightening their bone mass and making them a bit more mobile. Sauropods have been found on every continent.

Sauropods are commonly found without a skull for at least two reasons. First, the skull had a fairly weak attachment site on the neck that likely released during transport. Secondly, the skull bones are thinner than most other bones and more fragile and harder to preserve. This "missing skull" problem was the reason Professor Marsh placed the wrong head on his original *Brontosaurus* reconstruction. He merely used one from a few miles away that seemed to fit.

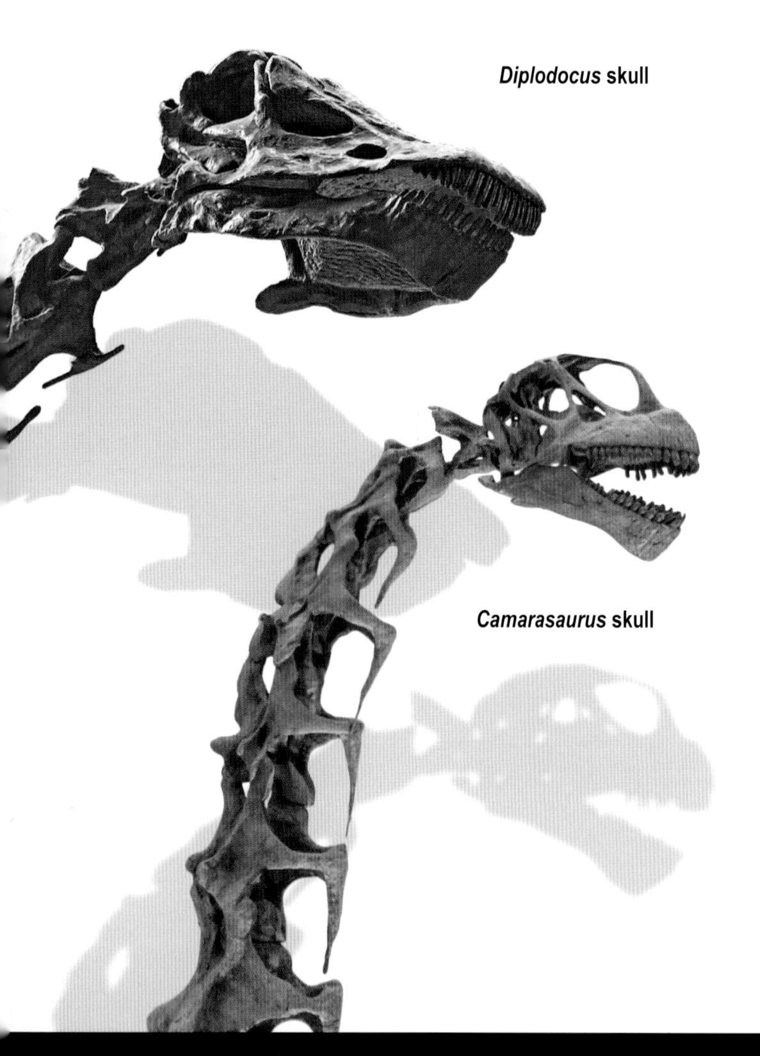

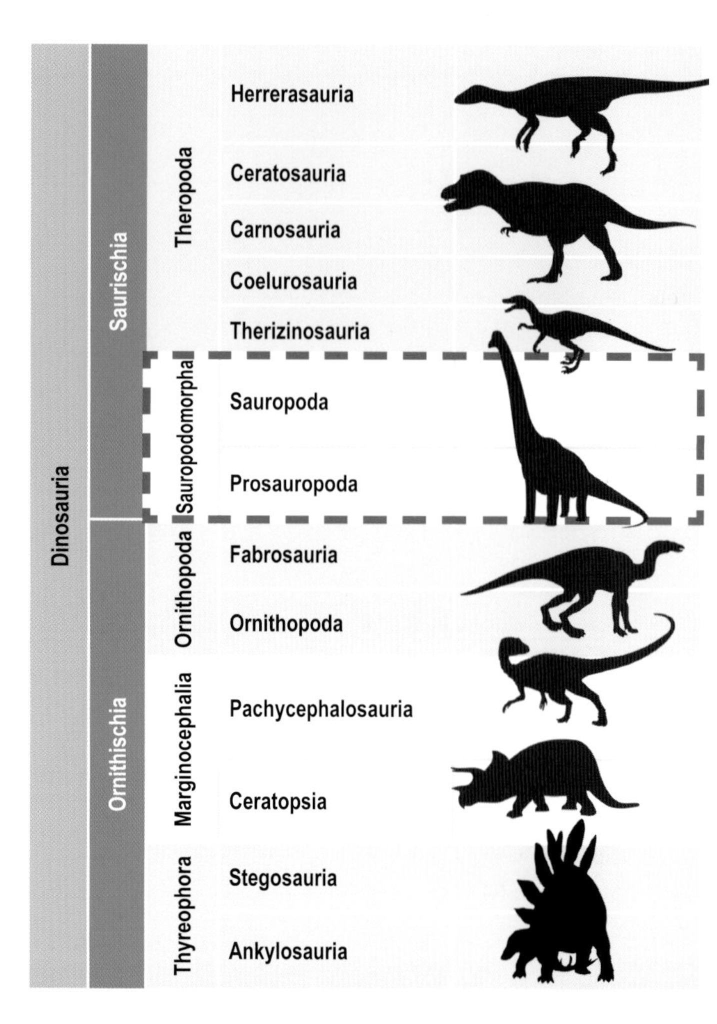

Another family of sauropods is the titanosaurs. Differences in the tail vertebrae distinguished these sauropods from the other families. However, these minor differences may be mere variations in the sauropod "kind." *Saltasaurus,* discovered in Argentina, was unusual because it had rows of bony plates embedded in the skin. Well-preserved embryotic skin inside the eggs of these dinosaurs showed the distinctive pattern for later bony plate development was present even in the earliest stages of life.

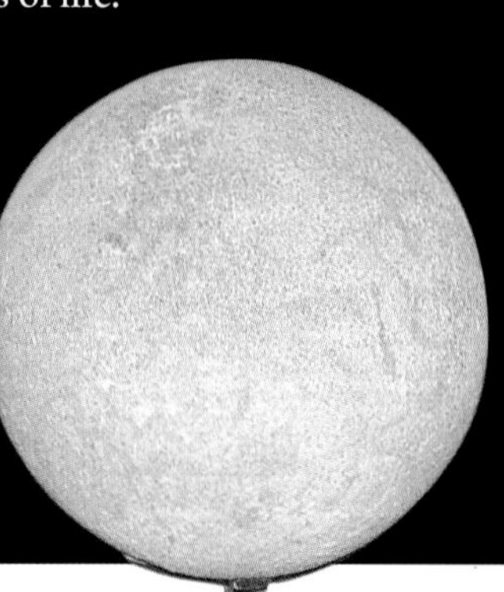

Photograph of a fossil cast of a *Saltasaurus loricatus* egg taken at the North American Museum of Ancient Life.

Brachiosaurus

Brachiosaurs are another famous group of sauropods. These dinosaurs were different than other sauropods because they had longer front legs than hind legs. They also had a large bump on the top of the skull. Up until recently, these dinosaurs were thought to be the heaviest. However, the discovery of *Argentinosaurus,* in Patagonia, Argentina, pushed the weight record to an estimated 70 tons, with a length of 115 feet. A partial skeleton of an even bigger sauropod from Argentina has been estimated at over 160 feet long. It is possible that God was describing one of these larger types of sauropods in Job chapter 40 as Behemoth.

GOD'S DESIGN IN THE SAUROPOD NECK

In 1809, Jean Baptiste Lamarck speculated that the necks of giraffes could grow longer simply by reaching higher and that they could somehow pass this trait on to their offspring. A half-century later, Charles Darwin suggested that giraffes born with longer necks than their rivals could reach more food and were better "fit" to survive. If Lamarck or Darwin were correct, scientists should find ample examples of ancient "giraffe" fossils reflecting a clear pattern of transformational neck growth over time. However, after generations of searching, paleontologists have failed to uncover any evidence of these transitional giraffe necks. It seems giraffes have always been giraffes.

What about the long-neck dinosaurs — do they show neck growth evolution? Recent studies of sauropodomorph dinosaurs have unlocked secrets to their marvelous design, but no evidence of evolution.[1] Sauropodomorphs are commonly referred to as the "long-necks" because all possessed a minimum of ten elongated neck vertebrae, making their necks at least six times longer than the tallest giraffe neck. By contrast, most mammals have only seven neck vertebrae. Manatees and sloths are the only exceptions. Three-toed sloths have eight to ten vertebrae, with nine being the most common. Two-toed sloths have five to eight vertebrae, with six being the most common. Manatees have six vertebrae.

Diplodocus skeleton.

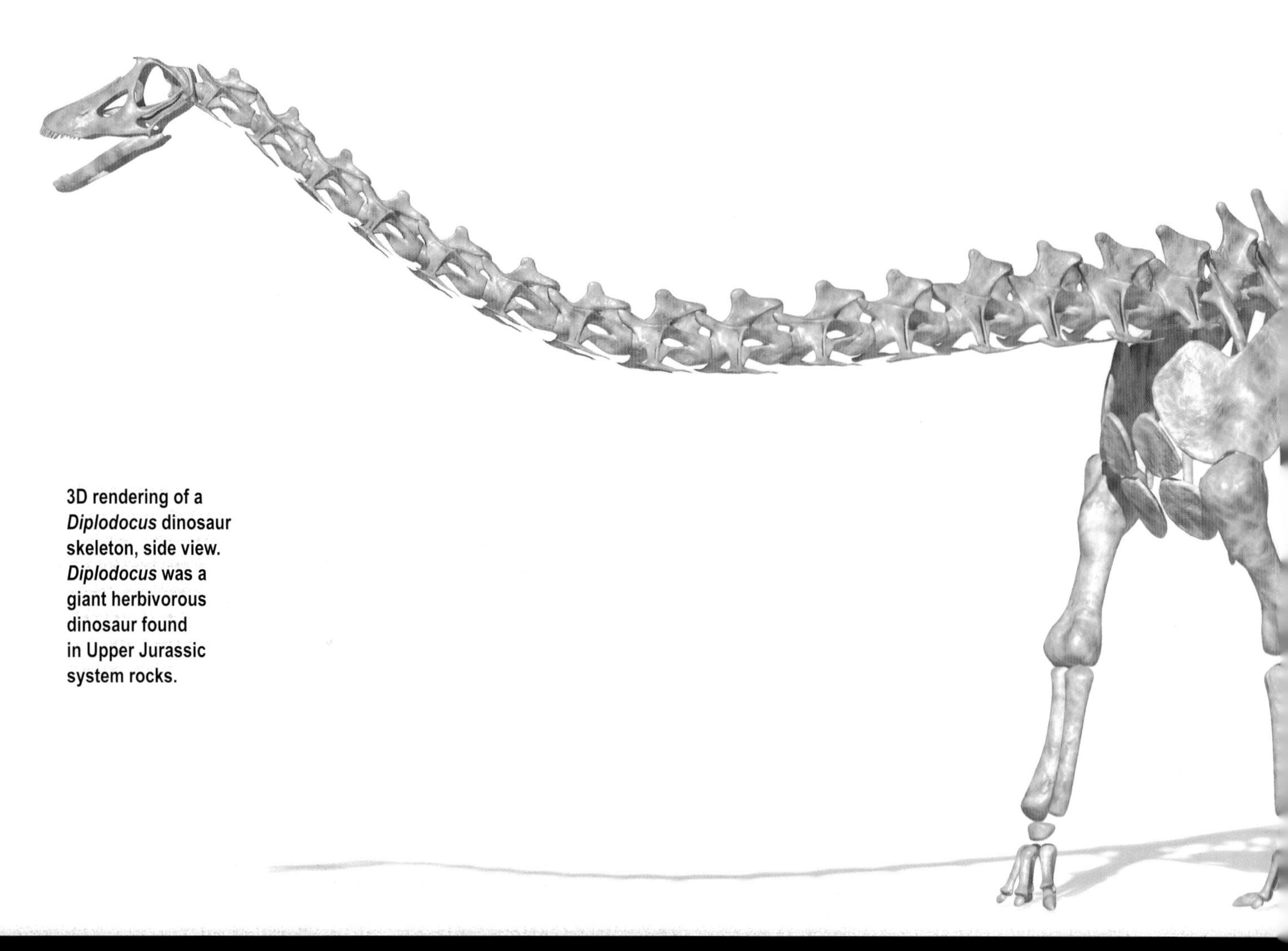

3D rendering of a *Diplodocus* dinosaur skeleton, side view. *Diplodocus* was a giant herbivorous dinosaur found in Upper Jurassic system rocks.

Secular paleontologists have claimed that sauropods "inherited long necks from their basal sauropod ancestors," but they offered no transitional fossils to back their claim. Instead they marveled at the complexity of the sauropod necks, noting that some were as long as 48 feet in length. These same scientists concluded that sauropods possessed a set of seven unique characteristics that facilitated an extremely long neck:

1. they had sufficient body size to support a long neck,
2. they had skeletal stability through a quadrupedal stance,
3. they all had small heads that were easily supported by the neck,
4. they all had ten or more neck vertebrae,
5. they had elongated vertebrae,
6. they may have had an air-sac breathing system, and
7. they had vertebrae designed with "pneumaticity," that is, many holes to lessen the overall weight of the neck and make it easier to hold up in the air.[2]

Although there is no absolute proof for an air-sac system in sauropods, the possession of a unidirectional air-sac system would have helped lessen the problem of "dead space" in the trachea when breathing through a long neck. This is caused when used air is re-inhaled before it can be fully exhaled — a limiting factor in the length of mammal necks. By allowing only one direction of air flow through the lung system, as in extant birds, this type of system always keeps freshly inhaled air passing across the lungs and eliminates the leftover, stale air that collects in mammals.

Without the perfect combination of all seven neck characteristics, the sauropods could not have survived. A large body size and stance are essential foundations for a long neck because they make a stable platform for long appendages and literally prevent the animal from toppling. The small heads of sauropods reduced the required lifting power of the necks. Take away just one characteristic and the sauropod would fall flat and fail as an organism. Each of these features would have to change simultaneously as the length of the neck grew. It is this all-inclusive, concurrent combination of alterations that makes any form of intermediate creature impossible. These animals simply could

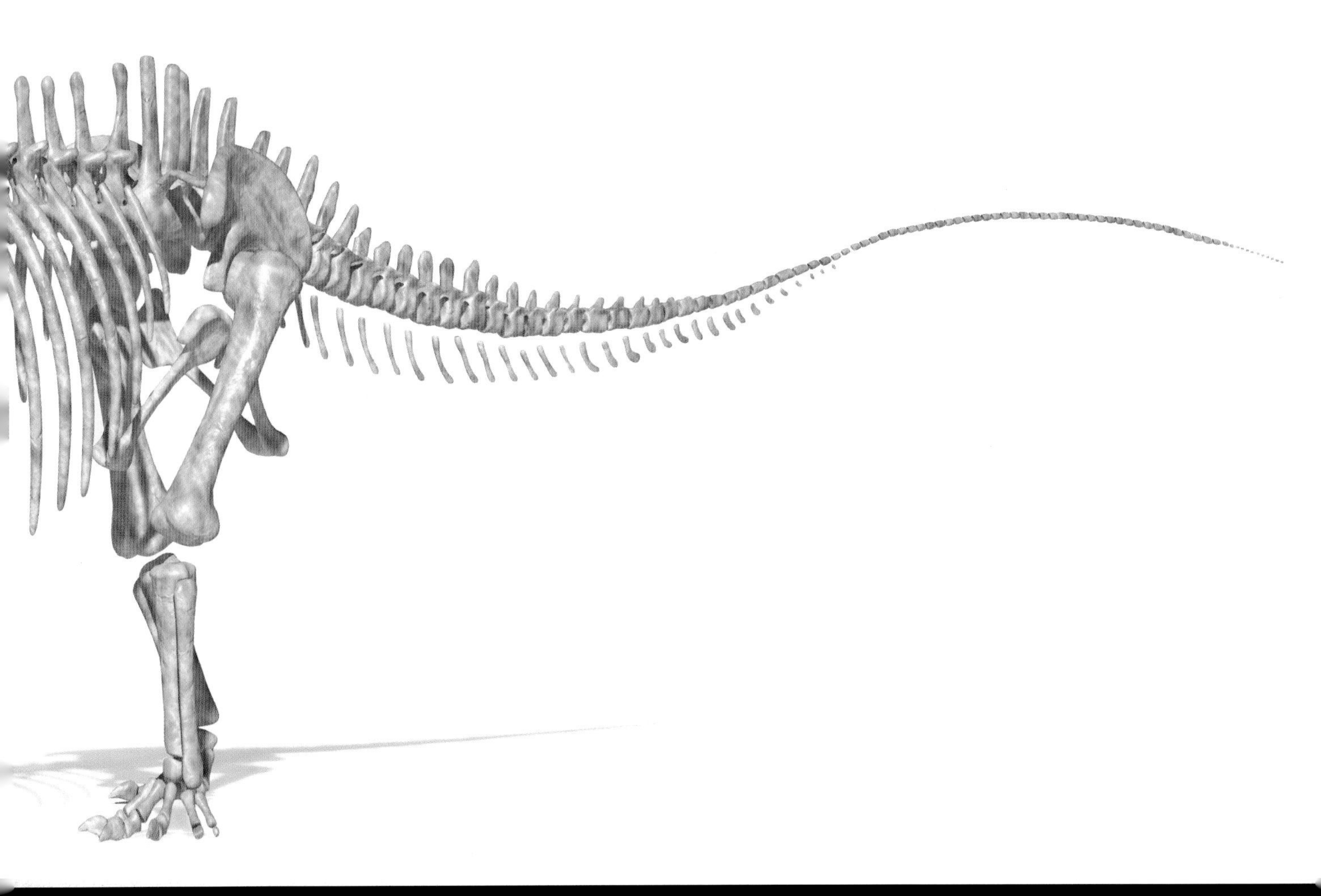

not have existed without possessing all of these characteristics, simultaneously. To suggest this perfect combination of characteristics appeared, fully functional, and by mere chance, is folly, especially since there are no known fossil ancestors to the Sauropodomorphs.

Overall, sauropods possessed the perfect combination of characteristics to have extremely long necks. It's as if they were designed that way. The Creator made fully formed sauropods, along with Man — and giraffes — on day 6 of creation week.

Professor Charles Gilmore of the Smithsonian Institution posing with caudal vertebrae (tail) assigned to *Diplodocus longus*, from specimen USNM 10865. Gilmore described the specimen and mounted it in the Smithsonian in 1932.

INFRAORDER SAUROPODA: *BRONTOSAURUS* OR *APATOSAURUS*?

O.C. Marsh made his own share of mistakes in his hurry to publish and name dinosaurs. The most famous is the controversy over *Brontosaurus* that became obvious with the issuing of four dinosaur stamps. In 1989, the U.S. Postal Service created stamps that included *Tyrannosaurus*, *Stegosaurus*, *Brontosaurus*, and *Pteranodon*. Many concerned citizens wrote to the USPS complaining that the name *Brontosaurus* should not have been used, as priority in science goes to the first name given to any species. Because the two names are synonymous and *Apatosaurus* was named first, they explained that the name *Apatosaurus* should have been used on the stamp instead.

USPS dinosaur stamps issued in 1989.

Before discussing the outcome of this controversy, we need to look at the history of these dinosaur discoveries. In 1877, Professor Marsh named the genus *Apatosaurus* (meaning deceptive lizard) on the basis of a vertebral column, giving a two-paragraph description with no illustrations in publication. Then in 1879, Marsh named the *Brontosaurus* (thunder lizard) from a pelvis and a few vertebrae, again with a two-paragraph description and no illustrations. Marsh went on to construct a full *Brontosaurus* skeleton in 1883 and, unknowingly, used the head of a *Camarasaurus* by mistake. *Brontosaurus* went on to become extremely popular as the Sinclair Oil symbol and even made several "appearances" in movies, including Disney's *Fantasia*.

Professor Marsh placed the head of a nearby *Camarasaurus* on his reconstructed *Brontosaurus* skeleton in 1883. Marsh simply grabbed the closest head he could find. Due to the popularity of the *Brontosaurus*, and the number of museums with copies, correcting this mistake took over 80 years to complete.

Apatosaurus

Mounted baby *Apatosaurus* in the Oklahoma Museum of Natural History.

Apatosaurus parvus

In 1903, after Marsh had passed away, Elmer Riggs at the Field Columbian Museum in Chicago (the future Field Museum) issued a proclamation in a rather obscure journal that the *Apatosaurus* Marsh named was actually a juvenile *Brontosaurus* specimen. He concluded the name *Brontosaurus* should be dropped in favor of the first scientific name issued, leaving only the name *Apatosaurus*.

The U.S. Postal Service responded with Bulletin 21744 invoking the Plenary Powers Rule, which says that popularized names can be validated even if older designations have priority. In other words, if a name is popular, it can still be scientifically correct. *Brontosaurus* is as correct to use as *Apatosaurus* for this genus. The stamps and the name *Brontosaurus* were saved from destruction.

Ironically, there was not a single letter complaining that *Pteranodon* was not a true dinosaur and should not have been included as a "dinosaur" stamp. Apparently, people were just as confused as to the definition of a dinosaur back in 1989 as they are today.

The *Brontosaurus* v. *Apatosaurus* controversy was reignited recently by paleontologist Emanuel Tschopp and his colleagues when they suggested that *Brontosaurus* should be a separate species from *Apatosaurus* after all. They concluded that there are significant differences in the skeletons to justify the separation. Other scientists disagree with their conclusions and prefer keeping them combined, as Elmer Riggs suggested. Creation scientists view both dinosaurs as the same biblical "kind," however, regardless if they are classified as separate species again in the future. The differences between the two are really quite trivial. They are merely an example of the diversity God built into the animal kinds.

The name *Apatosaurus* means "deceptive lizard." Apato = deceptive, and saurus = lizard. Looking back and ahead at all the controversy involved with this dinosaur, it seems appropriate that Professor Marsh had the unexpected foresight to name this dinosaur "deceptive."

INFRAORDER SAUROPODA: *DIPLODOCUS*: A JULY 4TH DISCOVERY

On July 4, 1899, a team from the Carnegie Institute in Pittsburgh (now the Carnegie Museum) found a toe bone of a large sauropod dinosaur near Como Bluff, Wyoming. They kept digging and soon realized they had a nearly complete skeleton of an 84-foot *Diplodocus* they nicknamed "Dippy." They believed this was a new species of this lengthy sauropod, naming it *Diplodocus carnegii* after Andrew Carnegie, in honor of their benefactor who was sponsoring the expedition.

The first *Diplodocus* was named in 1878 by O.C. Marsh who studied a specimen from Cañon City, Colorado, collected by Samuel Williston and Benjamin Mudge. Both this specimen and Dippy were found in the Morrison Formation, although in different states.

The Carnegie team quickly mounted the new *Diplodocus* specimen in their museum in Pittsburgh for the world to see. This was one of the first and longest dinosaurs ever placed on public display. The public demand was unprecedented. Andrew Carnegie had to hire a special team to make plaster copies of his new dinosaur to meet the need for requested copies. Over the next few years, replicas of this sauropod were distributed to most of the prominent museums of the world, from Europe to South America. This one dinosaur is often credited with jumpstarting the popularity of "dinosaurs."

There are four accepted species of *Diplodocus*. The first was called *D. longus*. All of the specimens are essentially the same, with only slight differences. Creation scientists consider these all versions of the same "kind" of animal, somewhat like breeds of dogs. It is likely that the sauropod dinosaurs like the *Mamenchisaurus*, from China, and the *Seismosaurus*, from New Mexico, are variations of this same "kind" also.

The name *Diplodocus* means "double beam," because this sauropod had two parallel rods arranged in a mirror-image fashion as chevrons along the underside of the tail. The tail was also composed of at least 80 separate vertebrae. The last 30 vertebrae were merely thinner and thinner rods, creating a whip-like end to the tail. Some paleontologists have suggested these tails could be rapidly "snapped," creating a potential cracking sound for warning the herd.

Diplodocids were rather slender for sauropods and weight estimates range from 10 to 18 tons. By contrast, *Brachiosaurus* probably weighed closer to 40 to 50 tons. Diplodocid limb bones were more slender than many other sauropods, and the ribs were long and narrow.

***Diplodocus* in the Berlin Natural History Museum.**

Carnegie Museum *Diplodocus* skeletal specimen

It is not unusual for the skulls of sauropod dinosaurs to be found missing from the skeleton. Evolutionary scientists think this is a consequence of water transport before burial, where the head, which wasn't well attached to begin with, became separated from the body. It is fairly common in paleontology to find the entire neck and body, and then to not find the head. Creation scientists agree with this interpretation to some respect, except they see the water transport as a consequence of the great Flood and not just some local flood event.

The design of the *Diplodocus* teeth makes this animal perfectly suited for eating soft plants and/or stripping leaves and needles from branches. Diplodocids had only peg-shaped front teeth, top and bottom, with no molars to chew with. They likely relied on gastroliths (so-called stomach stones) in a special gizzard in order to digest their food.

Brachiosaurus skeleton from Berlin Museum

Some paleontologists are placing *Brachiosaurus* into a new subcategory of sauropods called the Macronaria, which means "big noses." New category or not, all agree they are a bit different from most sauropods. These dinosaurs were reintroduced to the public in the movie *Jurassic Park* when a *Brachiosaurus* sneezed on a girl hiding in the trees and covered her with dinosaur "snot." Because of the large nasal opening on the top of *Brachiosaurus* skulls, early paleontologists believed these dinosaurs lived submerged in water, buoying their massive bodies and breathing out of the barely exposed tips of their skulls. More recently, studies of their compact foot structure suggest that they lived on dry land like most dinosaurs. Dr. Larry Witmer has even discovered that the nose openings on these dinosaurs were closer to the front of the skull than originally thought, and that the large openings were filled with fleshy nasal tissue. This design may have allowed an increased sense of smell in these dinosaurs.

Brachiosaurus altithorax was named by Elmer Riggs while at the University of Chicago in 1903. He had collected the type specimen in 1900 from a site in western Colorado. They recovered about 20 percent of the specimen, including some of the leg bones, ribs, pelvis, and vertebrae. There was just enough material to name the new species the "arm lizard," in honor of its long front legs. Most of this original type specimen is now stashed away and out of public view at the Field Museum in Chicago. The public can still view a replica outside the Field Museum, and another replica is located inside Chicago's O'Hare International Airport terminal.

A more complete specimen of a *Brachiosaurus* was recovered by the German expeditions to Tanzania in 1909–12 and described by Werner Janensch in 1914. Bone replicas from this more complete specimen allowed American paleontologists to construct and mount their original *Brachiosaurus* specimen. Recently the African specimen, housed at the Berlin Museum, was renamed *Giraffatitan brancai* due to slight differences in bone proportions, although not all paleontologists agree on this change. Creation scientists, by contrast, consider both of these brachiosaurs the same "kind," and any minor differences merely variations within God's design.

Surprisingly, Alfred Wegener never mentioned the codiscovery of brachiosaurs in North America and in Africa as evidence to support his fledgling hypothesis of continental drift. He mentioned other fossils that "matched" when he reconstructed his visionary Pangaea, but seemed to miss this opportunity to further boost his idea with dinosaur support. Other paleontologists of the day must have wondered how the same dinosaur could have coexisted on two continents now separated by thousands of miles of ocean. It wasn't until the modern theory of plate tectonics that this mystery was resolved.

Close up of *Brachiosaurus* skull in the Berlin museum. The large bulbous nose cavity on top may have served as a resonance chamber for the generation of sounds.

Another Macronarian dinosaur was found in the African nation of Niger by Paul Sereno. He named this dinosaur *Jobaria*, meaning "mythical animal." He found close to 95 percent of the type specimen, along with a juvenile of the same species, but, as usual, neither had a skull. He finally was able to locate a seemingly correct skull miles away with the help of a local nomad. It's possible this is the wrong skull like the original skull of *Brontosaurus*. The adult *Jobaria* was about 70 feet long and weighed about 25 tons.

Brachiosaurus is still one of the heaviest and strangest sauropod dinosaurs. It is estimated to have weighed up to 50 tons and had a length of 85 feet. The oddest feature is not the bulbous nose, but the longer front legs compared to the hind legs. This caused the animal to hold its head up extremely high in the air, possibly up to 60 feet above the ground! Paleontologists are still struggling trying to explain the blood pressures necessary to reach that height and supply the brain. God's design of this dinosaur was truly unique.

Parasaurolophus cyrtocristatus in the Evolving Planet exhibit at the Field Museum

CHAPTER 7: SUBORDER ORNITHOPODA: THE DUCK-BILLS

We now begin our survey of the Ornithischia, the bird-hipped dinosaurs. These dinosaurs had a completely different arrangement to the three pelvic bones compared to the Saurischia. The Ornithischia didn't have the pubis bone pointed forward and down in support of the abdominal region like the lizard-hipped dinosaurs. Instead, it had a pre-pubis bone pointing forward and in line with the vertebrae. Ornithischia dinosaurs had all three of their pelvic bones subparallel to the backbone.

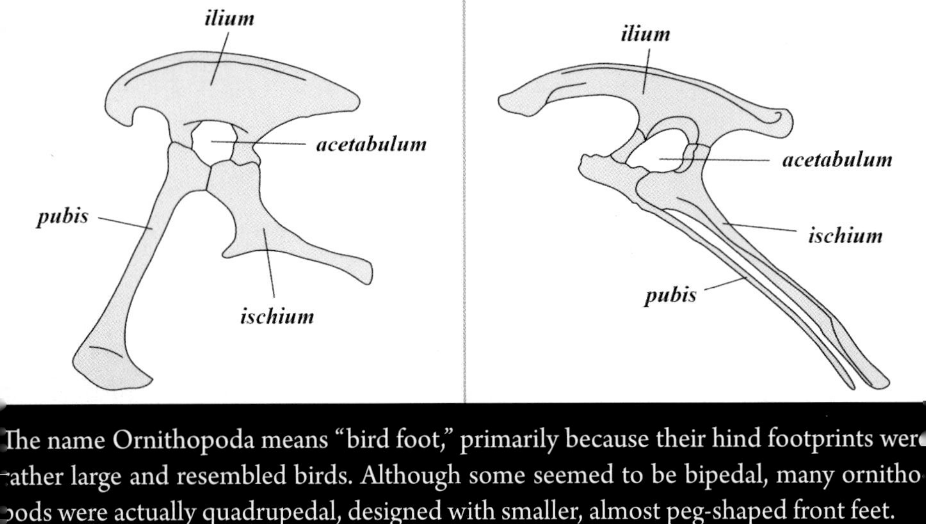

The name Ornithopoda means "bird foot," primarily because their hind footprints were rather large and resembled birds. Although some seemed to be bipedal, many ornithopods were actually quadrupedal, designed with smaller, almost peg-shaped front feet.

There are five recognized families of ornithopods. All of them were designed with their front teeth or toothless beak (premaxillary) set in the jaw a bit lower (or ventrally offset) than the cheek (maxillary) teeth. The lowering of the front teeth gave this suborder the "duck-billed" look. Three of the families that compose this suborder, the Heterodontosauridae, the Hypsilophodontidae, and Dryosauridae, were primarily bipedal with long shin bones on their hind legs, likely making them fast runners. They were rather small, usually less than 10 feet in length, and only varied from one another in minor skeletal differences, mostly in their teeth (possibly variations within the "kind").

***Heterodontosaurus* skull**

Heterodontosauridae were designed with front teeth, cheek teeth, and strange, canine-like tusks in between. There is a lot of debate about the purpose of these tusks. Some paleontologists suggest they were for display and/or were only on male members of the species. However, there is little data to back this claim. Others suggest these "herbivore" dinosaurs may have been omnivores or even meat-eaters, as the canines are present even in the hatchlings. These dinosaurs may have been one of the few groups that did not replace their teeth throughout their lifetime, like most dinosaurs. CT scans have shown no new teeth forming in the jaws of fossil specimens. Scientists note that even their molar teeth were odd, and more closely resembled those of mammals.

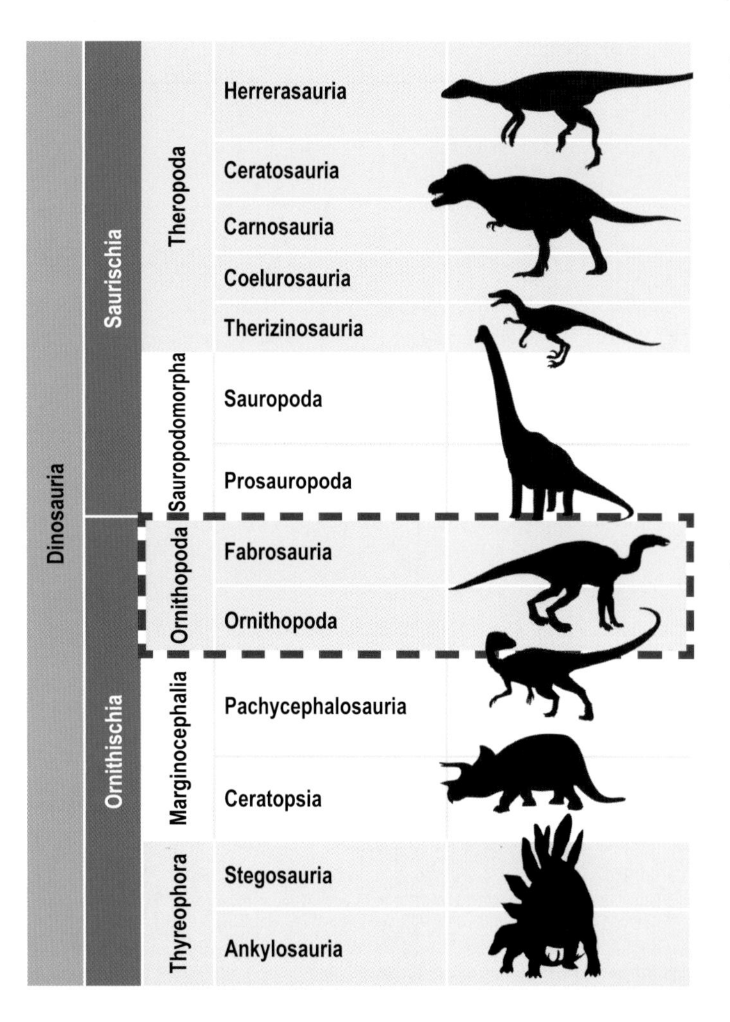

***Iguanodon* hand showing the proper attachment for the thumb spike.**

Dryosaurus was a 10 to 13-foot bipedal dinosaur found in Jurassic system rocks (late Flood) in Western USA and Tanzania. Oddly enough, the name literally translates as Oak Lizard. This genus was used as the namesake for the Dryosauridae family and resembled the Hypsilophodontids, except they didn't possess premaxillary or front teeth.

The most famous families of Ornithopods are the Iguanodontia and the Hadrosauridae. Most *iguanodons* were created with a "thumb" spike, whereas the hadrosaurs had a normal-sized first digit (thumb) on their hands. These two families of dinosaurs are mostly found in Cretaceous system rocks (late Flood). Both possessed a hardened beak or bill devoid of front or premaxillary teeth, but compensated with plenty of cheek teeth. These dinosaurs were much bigger, often 20 to 30 feet long or more, compared to many of the ornithopod families. As they matured, ossified tendons developed along the backbone, making the back and part of the tail stiff and straight. This additional support feature helped to support the adult weight of the dinosaurs like struts. Most dinosaurs in these two families walked as quadrupeds, but some may have run on just their hind legs.

Hadrosaur cheek teeth were plentiful, and there were at least three replacement teeth below each active tooth position. The teeth wore to a sharpened blade-like shape at the top, forming a cutting surface for slicing vegetation with a scissor-like motion. The long rows of teeth could process very coarse material. Stomachs of mummified hadrosaur skeletons contain twigs, conifer needles, and coarse seeds.

The family Hadrosauridae is further divided into two subfamilies, the Lambeosaurines and the Hadrosaurines. The Hadrosaurines had the flatter "Roman nosed" nasal area, while the Lambeosaurines had crests and tubes on the top of their heads that were connected to their nasal passages.

Lambeosaurines were named after the Canadian paleontologist Lawrence Morris Lambe who was born in 1863. He searched the dinosaur beds in western Canada in the late 19th century and early 20th century with the

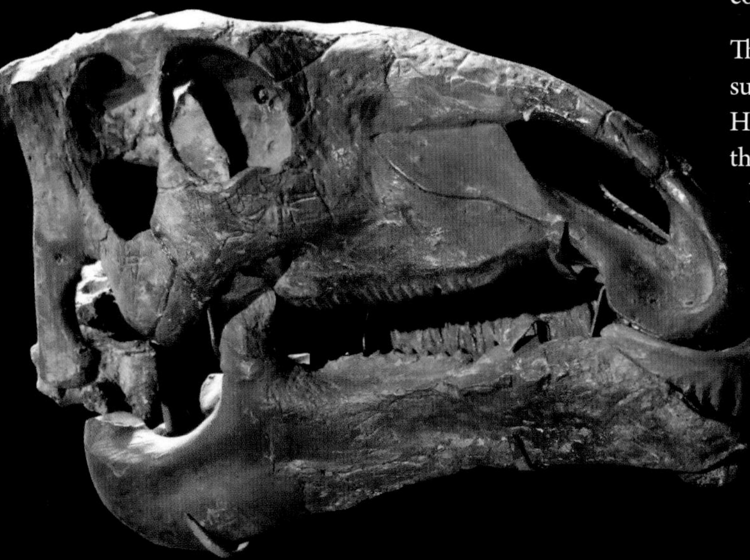

Hadrosaurine dinosaur with no head crest.

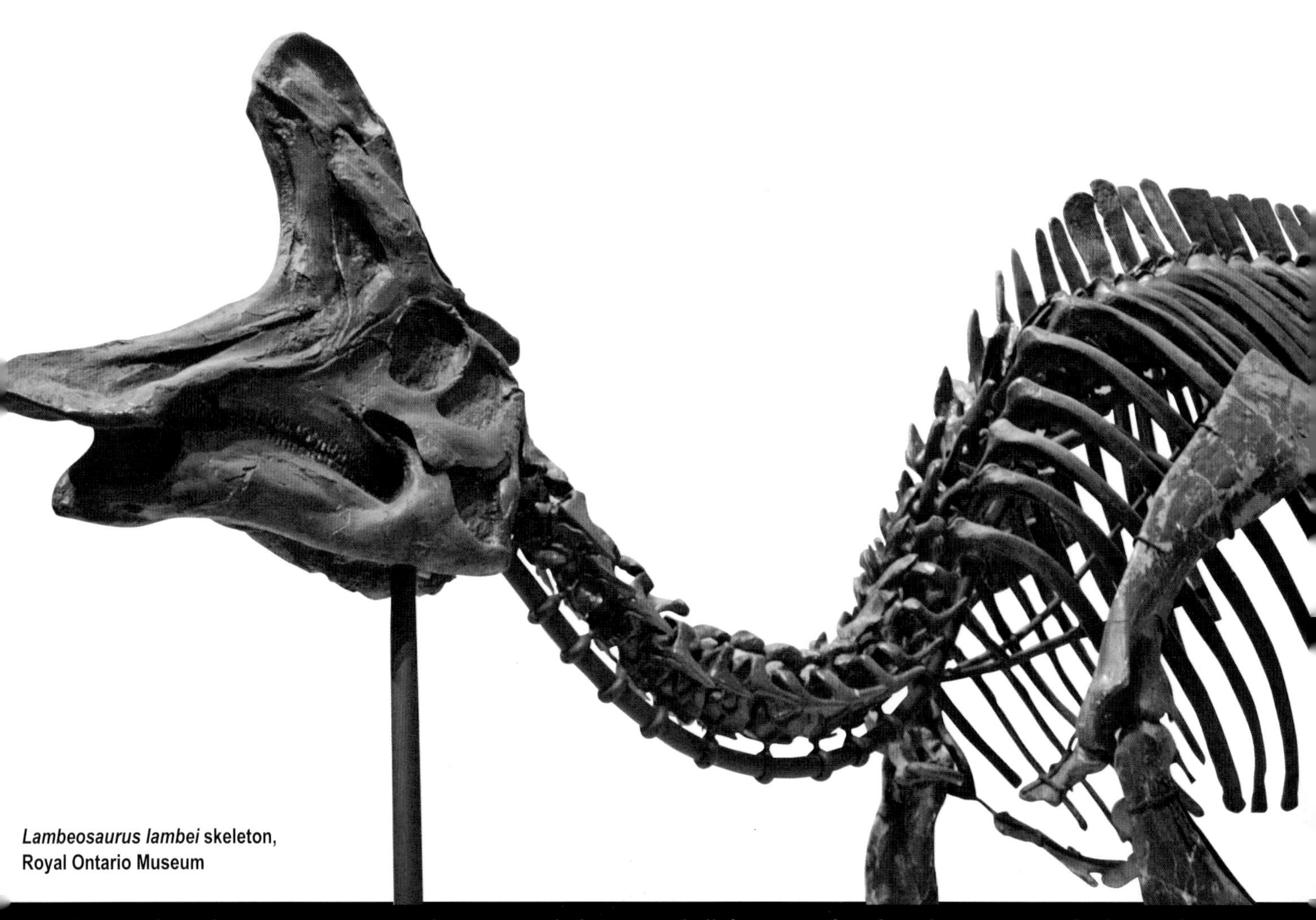

Lambeosaurus lambei skeleton, Royal Ontario Museum

Canadian Geological Survey, naming many dinosaurs, including *Edmontosaurus* (1917), *Styracosaurus* (1913), and *Gorgosaurus* (1914). He also named and described fossil crocodiles, fish, and corals. He passed away in 1919. *Lambeosaurus,* a crested hadrosaur, was named in his honor in 1923.

Maiasaura, a type of Hadrosaurine, was named by John R. Horner for its perceived "mothering" of its young hatchlings. Horner presented evidence suggesting that *Maiasaura* didn't possess fully developed limb bones upon hatching and could not immediately leave the nest. He also proposed that the crushed eggshell fragments found in the nests were further evidence the hatchlings stayed in the nest after hatching, stepping on and breaking the egg shells. Creation scientists explain these baby dinosaur fossils as animals that hatched in the Flood year, surviving a few days in the nest, before the Flood waters buried them rapidly in sediment, nest and all.

ICR's "Eddie" the *Edmontosaurus,* discussed in chapter 1, is also a hadrosaurine like the *Maiasaura.*

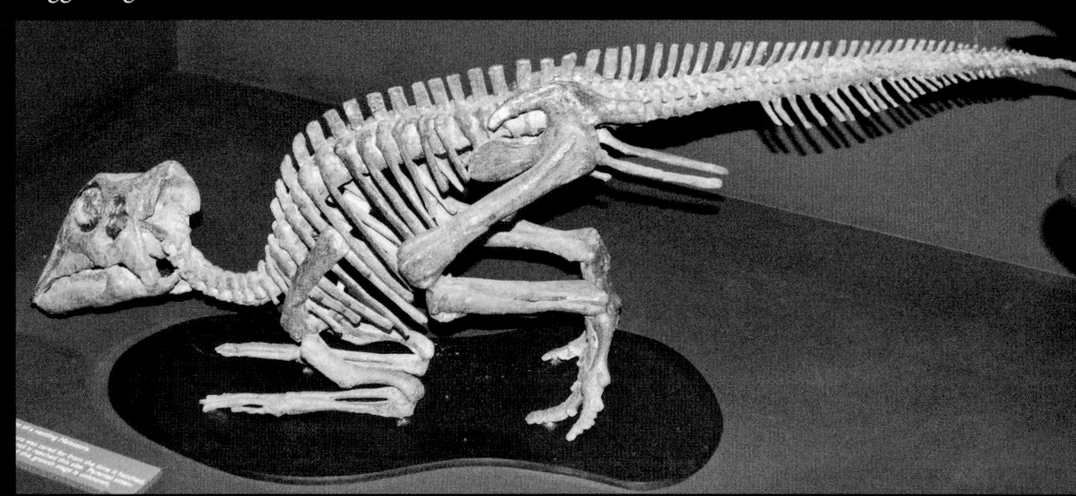

Maiasaura **hatchling on display at the Museum of the Rockies in Bozeman, Montana.**

INFRAORDER ORNITHOPODA: *PARASAUROLOPHUS*: THE CRESTED REPTILE

Although possibly the same "kind" as *Maiasaura* and *Edmontosaurus*, *Parasaurolophus* is a member of the lambeosaurine subfamily. *Parasaurolophus* and many others, such as *Corythosaurus* and *Lambeosaurus*, possessed large nasal crests on the top of their heads. Named by William Parks in 1922, the name *Parasaurolophus* means "beside crested reptile." Most of the fossils of this species are found in western North America in Cretaceous system rocks that are considered late Flood by most creation scientists.

Parasaurolophus grew to about 40 feet long and weighed about two tons. They may have walked on all four legs, as some trackways seem to suggest, and ran on only their larger hind legs. *Parasaurolophus* likely lived in swampy, wet areas, as there is some indication that they had at least partially webbed-feet. They are commonly found with many other specimens of the same kind, so it seems likely that they traveled in large herds.

New lambeosaurine hatchlings haven't been found with any large crests on the tops of their heads. However, a yearling *Parasaurolophus* from Utah was discovered with a small bump on its head. Presumably, this small bump would grow with time into the six-foot-long crest observed on a full-grown *Parasaurolophus*. Speculation about the use of these large crests began with the earliest discoveries of these dinosaurs. Early workers believed the tubes allowed the dinosaurs to stay submerged in water and use the crests as snorkels. After noting there were no holes in the top of the crests, paleontologists now think these features were primarily for sound-making, and possibly for sexual display, somewhat like deer antlers. Scientists have blown air through these tubes and suggested they were used for making sounds, like a trombone, for signaling danger, distress, group recognition, and/or mating calls. The diversity in the shapes of the crests would have provided a different sound for each family unit, similar to different brass instruments.

Hadrosaur specimens like "Leonardo" and "Dakota" are sometimes found as dinosaur mummies. These are not mummies like we normally think, but are specimens with intact skin imprints wrapped around an articulated dinosaur skeleton and sometimes organ imprints too. George and Levi Sternberg found the first "mummified" hadrosaur in Wyoming in 1908. Since that time, many more have been unearthed, including a *Corythosaurus* and even a horned dinosaur, *Monoclonius*. These specimens show all duck-billed dinosaurs had scaly, reptile-like skin, from the juveniles to the adults.

Parasaurolophus specimen with a pathogenic v-shaped notch in spine.

Parasaurolophus

One spectacular mummified hadrosaur was discovered in North Dakota in 1999. In addition to skin prints, this fossil had well-preserved foot pads, keratin on its "hooved" feet, and actual rust-brown skin scales with intact cell structures. The CT scan of the skin hinted that a striped skin pattern may have been present along the hindquarters and tail. Scientists also reported finding fossil ligaments, tendons, and organs. The remarkable preservation of this specimen allowed scientists to determine it had more muscle mass in its hind legs than previously thought, making it possible to even outrun many predators, including *T. rex*. The well-preserved soft tissues also indicate that this dinosaur was not millions of years old, but only thousands. This dinosaur was buried rapidly in water and sediment during the great Flood less than 4,500 years ago, preserving many of its fine details.

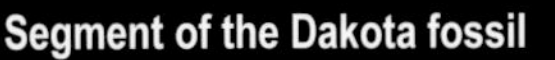

Segment of the Dakota fossil

Dakota skin impression

Dakota is the nickname given to a fossil *Edmontosaurus* found in the Hell Creek Formation in North Dakota. The fossil is unusual and scientifically valuable because soft tissue, including skin and muscle, have been fossilized, giving researchers the rare opportunity to study more than bones, as with most vertebrate fossils. Preliminary research results indicate that hadrosaurs had heavier tails and were able to run faster than was previously thought.

Triceratops specimen (nicknamed "Lane") at the Houston Museum of Natural Science.

CHAPTER 8: SUBORDER MARGINOCEPHALIA: THE DOMED, HORNED, AND FRILLED

The term Marginocephalia translates "edge head." All of the dinosaurs placed in this suborder were designed with some type of frill or ridge along the back of the skull. Most Marginocephalia fossils are found in Cretaceous system rocks, likely surviving until later in the Flood than many dinosaur groups. The dinosaurs in this group also possessed the bird-hip style of the Ornithischia.

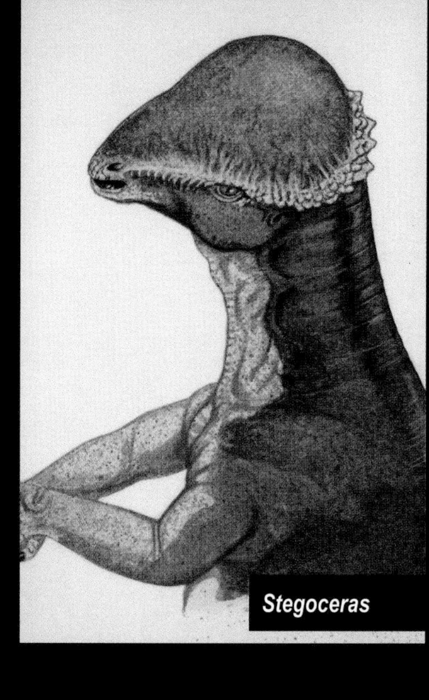

Stegoceras

This suborder has been divided into two rather diverse infraorders, the Pachycephalosauria and the Ceratopsia. Pachycephalosauria means "dome head," and Ceratopsia means "horn head." The observed diversity in the marginocephalia is due to tremendous variation within the created "kinds."

The Pachycephalosauria were bipedal, dome-headed dinosaurs, designed with a thickened bone on the roof of the skull. A common dinosaur from this group is the genus *Stegoceras* found in Western North America. Other Pachycephalosaurs have been found in Europe and Asia. Their fossil record is fragmentary at present.

Stegoceras, like all Pachycephalosauria, had small serrated teeth for eating and slicing vegetation. The skull was designed to attach to the neck at an angle to the backbone so that the head hung down and the top of the domed skull was out front and pointed forward. These dinosaurs had many similar body features in common with Ornithopods. They had much longer hind limbs (bipedal) and a tail that was strengthened with ossified tendons for support. Most paleontologists think the domed head was used for competition within the species, like bighorn sheep today. It may have been used as a defense when threatened, or simply for sexual or social display.

Pachycephalosaurus wyomingensis dinosaurs

Infraorder Ceratopsia is further divided into Psittacosauridae and Neoceratopsia. The Psittacosauridae is represented by one genus, *Psittacosaurus*, also known as the "parrot lizard."

Psittacosaurus was another bipedal, plant-eating dinosaur that was first associated with Ornithopods, but now has been aligned with the Ceratopsia. This dinosaur had a toothless beak, resembling the shape of a parrot, and wide, flat, self-sharpening cheek teeth for chewing coarse vegetation. The "frill" on the back of the skull is termed "rudimentary," as it is nearly nonexistent. It is more like a ridge or long bump along the back base of the skull.

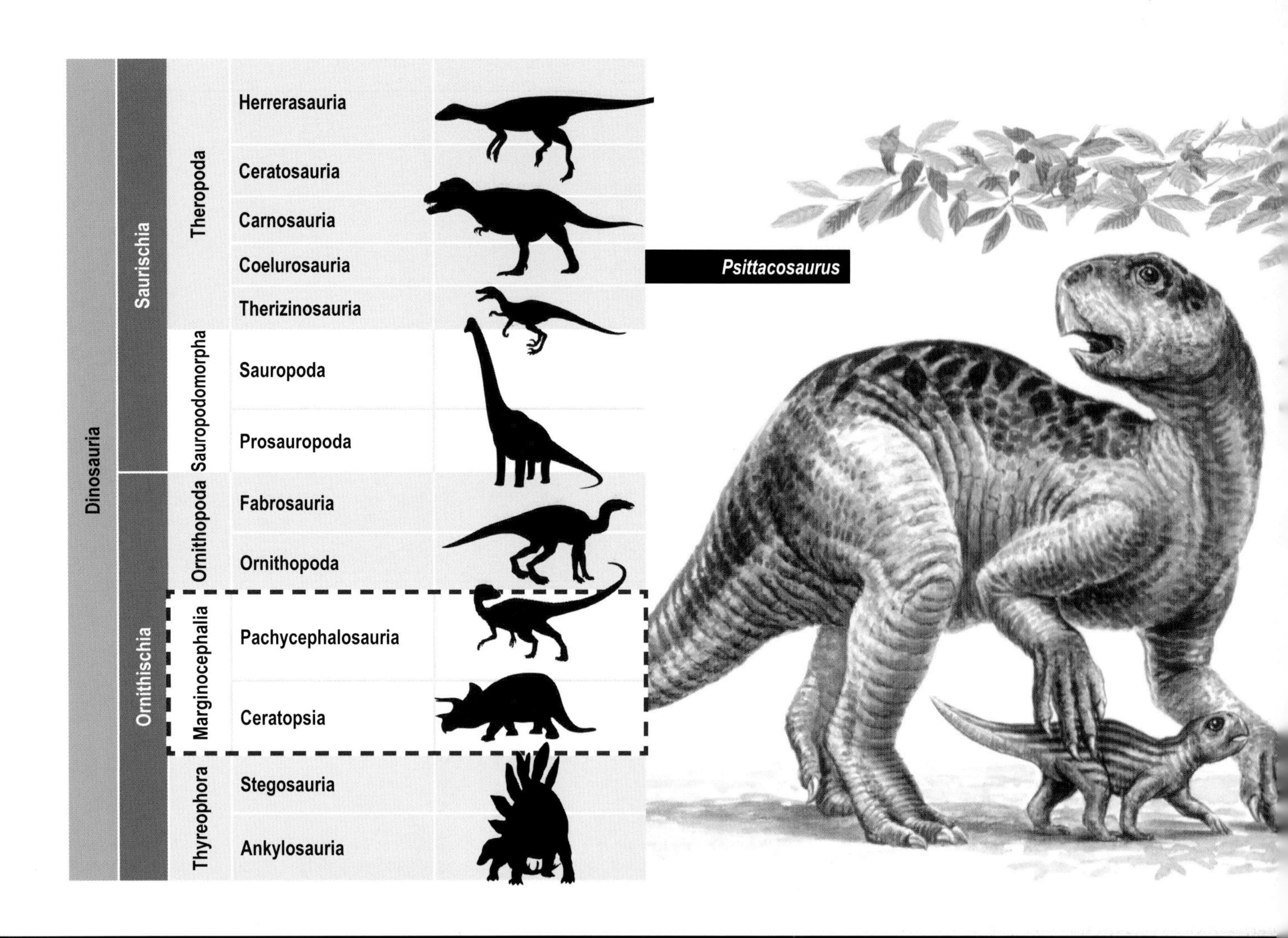

Neoceratosia includes all the famous horned and frilled dinosaurs such as *Triceratops*, *Styracosaurus*, and *Centrosaurus* from North America. Neoceratosia also includes the Protoceratopsids from North America, and primarily Asia, and which had no pronounced horns on the head. Ceratopsia is one of the most diverse groups of dinosaurs in terms of the variety of horns, frill styles, and shapes. Neoceratopians were all designed as obligate quadrupeds.

Most adult Ceratopsia were quite large at 20 to 30 feet long. They were designed with extremely large skulls (up to 8 feet long), that exhibited wide differences in horn styles and frill shapes. The basic body structure of each variety was the same design, however, and even looked identical as hatchlings. It wasn't until they matured that the differences in horns and frill shapes became evident. They had a toothless, beaked jaw with brackets of cheek teeth similar to hadrosaurs, including at least 3 or 4 replacement teeth beneath the active teeth. The teeth came together at a high angle, producing a scissor-like action. These dinosaurs could probably chew up the coarsest plant material because their frills provided attachment sites for large jaw muscles.

Many paleontologists used to think the frill was used for protection from predation, but after careful examination, it seems less likely. First, the frills usually had large openings in them, which do not offer much protection. Secondly, the frills are highly vascularized and filled with internal openings, similar to stegosaur plates, making them likely to bleed if broken or bitten. Many paleontologists now think the frills were primarily for muscle attachment, giving a strong bite, and possibly for thermoregulation (heating and cooling by flushing with blood). Others think the frills served for family recognition and even sexual display.

***Protoceratops* skeleton.**

Ceratopsian frill

INFRAORDER CERATOPSIA: *TRICERATOPS*: JUST A YOUNG *TOROSAURUS* AFTER ALL?

Until recently, *Triceratops* was thought to be one of the only Ceratopsia that didn't have holes in its frill. However, after examining numerous specimens, John R. Horner and his colleagues proposed that *Triceratops* is merely an older juvenile version of *Torosaurus*.[1] He was able to demonstrate how the ontology or growth pattern of these genera merged into one another as they matured. Holes in the frill did develop in the older adults. It has also been shown that many Ceratopsian dinosaurs underwent changes in horn angle and frill ornamentation pattern as the dinosaurs grew and matured. Ontological changes are one of the reasons for so many different-looking dinosaurs, and likely, the misnaming of so many species. Many so-called dinosaur species are probably just examples of different growth stages and/or small variations within one created "kind."

Triceratops, or *Torosaurus*, was designed with a very large skull, up to 7 or 8 feet long. The first four vertebrae of the neck were fused to support the great weight of this massive head. The adult dinosaur was about 25 to 30 feet in total length. The tail was short and narrow and the pelvis was fused to the backbone, placing the tail in a slight downward position. The four legs of the dinosaur were of near equal length, suggesting great power and speed when it walked or ran. Footprint evidence suggests these dinosaurs walked at 4 to 8 mph and might have been able to run at 20 mph. A few paleontologists think they could have run up to 30 mph. Unfortunately, there is no fossil footprint evidence to back up these suggested higher speeds.

For decades there has been debate on the front leg posture of *Triceratops*. Most early reconstructions had the front legs sprawling in lizard fashion, and the hind legs erect, as in all dinosaurs. The debate seems to have been settled after the discovery of some Ceratopsian trackways in Colorado that indicate an upright posture for both front and hind limbs. Many museum specimens had to be reconstructed to reflect this "new" upright posture.

Triceratops was created with two large horns above the eyes and a smaller horn on the tip of the nose. These were different from antlers, as they were a permanent part of the skull and were made of solid bone. There was a sheath or covering on the horns of keratin similar to fingernail material that sharpened and slightly extended the length of the horns.

Originally, the horns of *Triceratops* were thought to be used in counterattack on

predators, like *T. rex*. However, more recent interpretations have the *Triceratops* and other Ceratopsians using their horns against one another in "territorial" or "mating" competitions. Horn scratch marks have been observed on the fronts of some frills that seem to support this horn-to-horn competition. However, it seems likely they would also have used their sharp horns as a defense against predators.

Ceratopsian dinosaurs appear to have traveled in great herds, as paleontologists often find them in widespread bone beds with hundreds of individuals of the same species. Creation scientists interpret these buried herds as evidence supporting the catastrophic nature of the great Flood. Ceratopsian dinosaurs were likely herded together, like other dinosaurs, trying to escape the rising Flood waters that eventually engulfed them.

A recent study suggested that a *T. rex* may have decapitated a *Triceratops* in the course of feeding on it.[2] Paleontologists found the skull separated from the body of the Ceratopsian along with numerous bite marks matching *T. rex* teeth on the frill and face of the *Triceratops* skull, suggesting it pulled the head off with its jaws while holding the body with its feet. This event likely happened while the Flood waters were actively rising, and these two dinosaurs were able to find some temporary high ground.

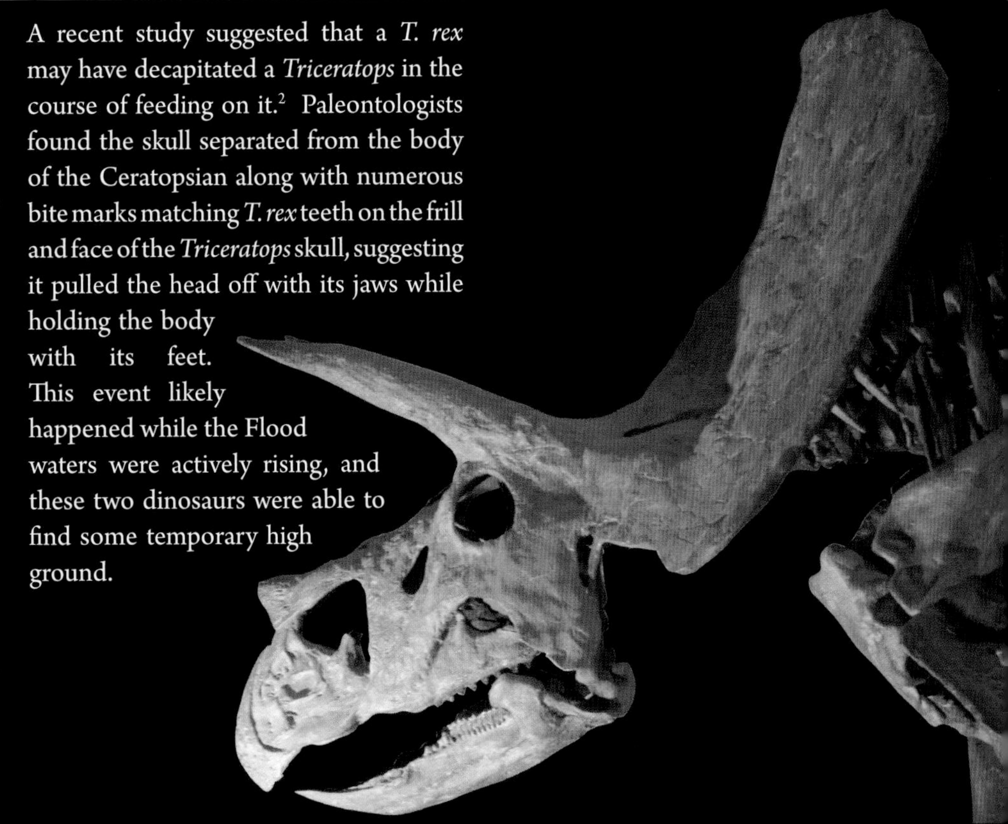

INFRAORDER CERATOPSIA: *PROTOCERATOPS*: THE "SHEEP" OF CERATOPSIAN DINOSAURS

The story of *Protoceratops* begins on the first Central Asiatic Expedition to Mongolia. On September 2, 1922, expedition photographer J.B. Shackelford picked up a skull of a dinosaur. He also noticed eggshell fragments nearby that were thought to be from a prehistoric bird. Walter Granger and W.K. Gregory described this new dinosaur in 1923 and called it *Protoceratops andrewsi* in honor of the expedition leader. The name means "first horned face." The following year, the second Asiatic expedition found egg nests containing up to 13 eggs each. Because they found 70 skulls and 14 skeletons of *Protoceratops* in the same layers as the eggs, they naturally assumed they were *Protoceratops* eggs. However, in 1993, an embryotic skeleton was found in a similar egg showing it was an *Oviraptor* egg and not a *Protoceratops* egg. The so-called egg-stealing dinosaur (*Oviraptor*) had become the "mother" dinosaur and the eggs are now attributed to *Oviraptor*.

Protoceratops was a sheep-sized dinosaur, standing 2 to 3 feet high, and about 5 or 6 feet long. It weighed a bit more than a sheep at about 400 pounds. Based on the number of skeletons found in single locations in Mongolia and China, evolutionary scientists think this dinosaur herded in large numbers, gaining the reputation as the sheep of dinosaurs.

Again, creation scientists explain part of this apparent herding as a consequence of catastrophic Flood activity, where tsunami-like waves pushed dinosaurs together in massive deposits, simulating herd behavior. Later expeditions by the American Museum of Natural History found lots of other animals mixed in the layers with the *Protoceratops* fossils, including crocodiles and turtles and even *Velociraptor* fossils.

Many *Protoceratops* skulls have sizable frills (half the length of the skull), whereas others had short compact frills on the back of the skull. Some paleontologists think these two sizes may be related to differences in the sexes. Some believe the larger frilled specimens are males and vice versa. As there is no clear way to distinguish the different sexes of most dinosaurs to date, this matter still remains unresolved.

One of the functions of the large frill design on Neoceratopsian dinosaurs was for the attachment of the jaw muscles. It seems that a larger frill would provide for longer muscles and a stronger bite. *Protoceratops* had no horns to protect itself with when attacked, and its frill had two large, open holes, called fenestra, one on either side. This dinosaur wasn't very large, only 4 to 8 feet long, but the large frill gave it a strong bite.

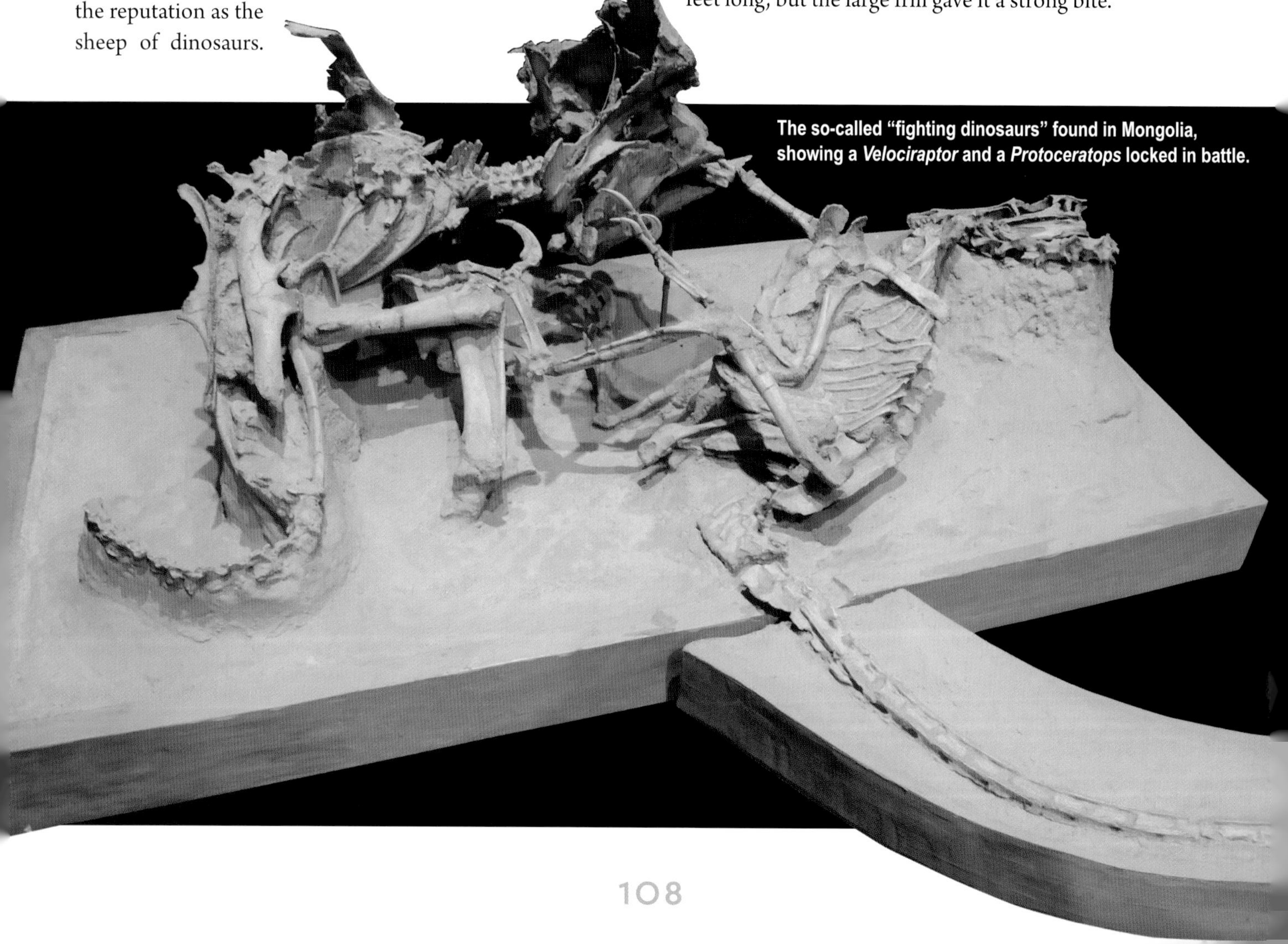

The so-called "fighting dinosaurs" found in Mongolia, showing a *Velociraptor* and a *Protoceratops* locked in battle.

One of the most spectacular fossils ever unearthed is the so-called "fighting dinosaurs" found in Mongolia, showing a *Velociraptor* and a *Protoceratops* locked in battle at the instant of burial during the great Flood. The *Protoceratops* has the leg of the predator in its jaws, preparing to bite, using its only counterattack, defensive measure. Whether it could have used its great jaw strength to snap the legs bones of the *Velociraptor* is unknown. We are only able to witness this struggle because the Flood sediments buried them both so rapidly, in life position, as a snapshot of history.

A nest of 15 young *Protoceratops* hatchlings were found clustered together in what many paleontologists believe represented a nest. Creation scientists explain the rapid burial of these hatchlings as a consequence of the Flood that engulfed the young dinosaurs soon after they hatched. During the Flood, many dinosaurs survived for some time, maybe many weeks, before they succumbed to the rising Flood waters. Dinosaurs that were pregnant at the onset of the Flood would have followed instincts and made nests and laid eggs on the increasingly limited dry land available. Because of the global nature of this catastrophe, we find dinosaur eggs, nests, young hatchlings, juveniles, and adults at every stage of growth all preserved as fossils in Flood deposits.

Greek writings as far back as 650 B.C. describe the mythical griffin, a four-legged lion-like animal with a raptor-like beak. One hypothesis ties the source of this creature to ancient nomads searching for gold across what is now Mongolia. These nomads are thought to have run across the exposed skulls of fossils like *Protoceratops*, and developed the griffin to explain what they saw as a consequence. Whether or not this is the true source of the mythical griffin is unclear. It is just as likely to have been a rendition of an actual living dinosaur or dragon in 650 B.C., instead of derived from fossils.

Griffin

Protoceratops

INFRAORDER CERATOPSIA: *PSITTACOSAURUS*: THE PARROT REPTILE

All Ceratopsian dinosaurs, and only ceratopsian dinosaurs, had an extra bone in the front of the jaw, called the rostral. This extra bone fit in between the normal premaxillary (front jaw) bones and is known as the "beak bone." Psittacosaurids had this bone too, and that's why they are classified with the Ceratopsians, which includes the great horned dinosaurs like *Triceratops*. The Psittacosaurids were a strange group. They didn't possess a frill on the back of the head like most Ceratopsians, nor did they have large horns.

Psittacosaurus is the best known dinosaur in this group. Its name means "parrot reptile," as it looks somewhat like a parrot in facial profile. In fact, its jaw was unusual for dinosaurs because it did not have a segmented jaw with internal jaw joints. Instead, it had a single upper jawbone and a single lower jawbone. The beak was rounded and flattened and not sharp, but seems to have been designed for cropping vegetation. Although it had no front teeth, it did have ample cheek teeth built for slicing, not grinding.

Psittacosaurus was first found in Mongolia on the first Central Asiatic Expedition led by Roy Chapman Andrews in 1922. Henry Fairfield Osborn, the head of the American Museum of Natural History, described and named the species in 1923. An adult *Psittacosaurus* was about 5 to 7 feet long and weighed about 50 pounds. Over 400 specimens of the psittacosaurids, and at least 10 species, have been found with sizes ranging from hatchlings to adults. All likely represent the same biblical "kind," as the differences are subtle. The psittacosaurids are found in later Flood rocks identified as Lower Cretaceous system strata.

Psittacosaurus had back legs much longer than its front legs. For this reason, paleontologists think it walked on two legs. The upper arms were fairly large and muscular, however, and possessed four fingers or digits, unlike the five digits found on most ceratopsians.

Although *Psittacosaurus* had cheek teeth, it didn't have the type of teeth to grind its food. It appears to have been an effective slicer, with teeth that self-sharpened as they wore. Like nearly all dinosaurs, they had ample replacement teeth beneath to replace worn-down teeth. The teeth were designed to chew coarse plants and possibly nuts. In order to digest vegetation, psittacosaurids had to rely on gastroliths, or stomach stones, to grind their food. Some have been found with more than 50 gastroliths in their abdominal region.

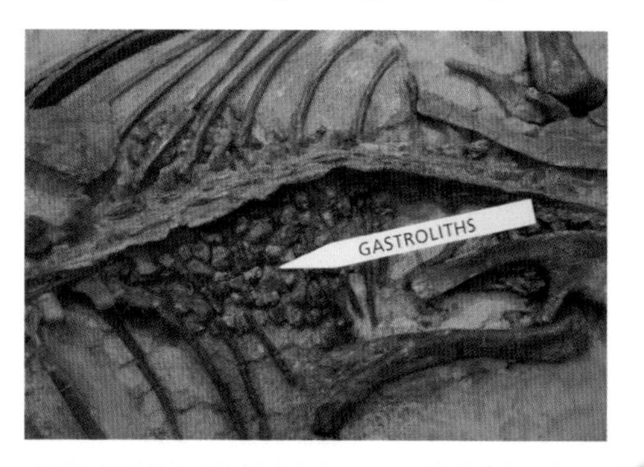

Psittacosaurus

Psittacosaurus nest, China. Adult skull was apparently added to the fossil during preparation.

An exceptional Chinese specimen of *Psittacosaurus* was found with what appears to be a "bristled" tail. Although skin imprints with scales were found across most of the specimen's body, there were 6- or 7-inch long, hollow, tubular bristles, in a single row, running down the top of the upper tail. Paleontologists aren't sure if the bristles were for display and/or social status or had some other use. Overall, the tail of *Psittacosaurus* was quite flexible, leading some paleontologists to suggest it was an aquatic or semiaquatic animal, using its tail like a crocodile, swinging it from side to side to assist in swimming. The bristled tail may have been designed to stick above the water, serving as a fin-like apparatus.

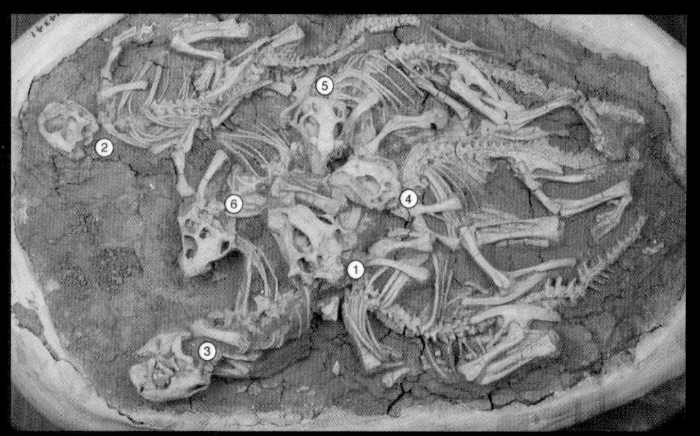

Cluster of six juvenile *Psittacosaurus* (IVPP V14341) from the Lower Cretaceous of Lujiatun, Liaoning Province, China. The specimen, illustrated as a photograph, shows six aligned juvenile specimens, of which specimens 2-6 are estimated to have been two years old at death, and specimen 1 was three years old, based on bone histological analysis.

Many hatchlings have been found of *Psittacosaurus*. These specimens had skulls less than two inches long and bodies about six inches long. One discovery in China found 32 babies with an adult skull nearby. Evolutionary scientists hailed this as proof of nurturing the young. They interpreted that a volcanic mudflow had buried the "mother" and the babies together. However, this discovery was later found to be another Chinese hoax. The adult skull was merely glued on to the nest of babies, making it look like the "mother." Creation scientists explain these buried and fossilized hatchlings as a consequence of rapid deposition by sediments transported by Flood waters. The newly hatched dinosaurs were engulfed before they had time to leave the nest.

INFRAORDER PACHYCEPHALOSAURIA: PACHYCEPHALOSAURS: THE DOME-HEADS

The first specimens of the infraorder Pachycephalosauria, or "dome headed reptiles," were found in the 1850s, but many of these were mislabeled due to the lack of fossil material. Most fossils are remnants of the thickened skull cap or dome and little else. Because of the fragmentary nature of these discoveries, some of these dinosaurs were even confused with the theropod *Troodon* because of their similar teeth.

Pachycephalosauria contains 11 species that have been found across the Northern Hemisphere, including North America, Europe, and Asia. Their fossils are found in later Flood rocks from strata of the Upper Cretaceous system. Pachycephalosaurs walked on two legs, with short upper arms, were about 6 to 7 feet long, and most weighed under 200 pounds. The design of these dinosaurs included the stiffening of the end of the tail with ossified tendons, reducing flexibility but adding support, similar to the Ornithopods and Dromaeosaurs. The tail likely functioned in a limited side to side motion as a counterweight for balance when walking and running.

The dome was commonly smooth on the top with pointy bumps sticking out around the back and sides of the dome. Most pachycephalosaurs also had a small ridge or shelf that developed around the back of the skull. This small ridge is why many paleontologists lump these dinosaurs into the suborder Marginocephalia ("ridge-headed") with the Ceratopsians like *Triceratops* and *Protoceratops*.

Pachycephalosaurus bones were first found by geologist Ferdinand V. Hayden in eastern Montana in 1859–1860. It wasn't until after Barnum Brown co-established the genus *Pachycephalosaurus* in 1943 that these earlier bones were identified as the same species. Unfortunately, even scientists today are not sure what a complete *Pachycephalosaurus* should look like, as most discoveries are just a skull cap and a few fragments. These dinosaurs may have been up to 15 feet long and weighed close to 900 pounds.

Stegoceras is one of the most common pachycephalosaurs. About 40 partial specimens have been found in Western North America in rock units like the Hell Creek Formation in Montana. This dinosaur was named and described by Canadian geologist Lawrence Lambe in 1902.

Pachycephalosaurs had some of the most diverse teeth of all dinosaurs. The teeth in the front of the upper jaw were cone-shaped, and the teeth on the sides of the jaws were leaf-shaped with ridges. Some species also had cone-shaped teeth in their lower jaw. The diversity of the teeth suggests these dinosaurs ate a wide variety of plants and vegetation.

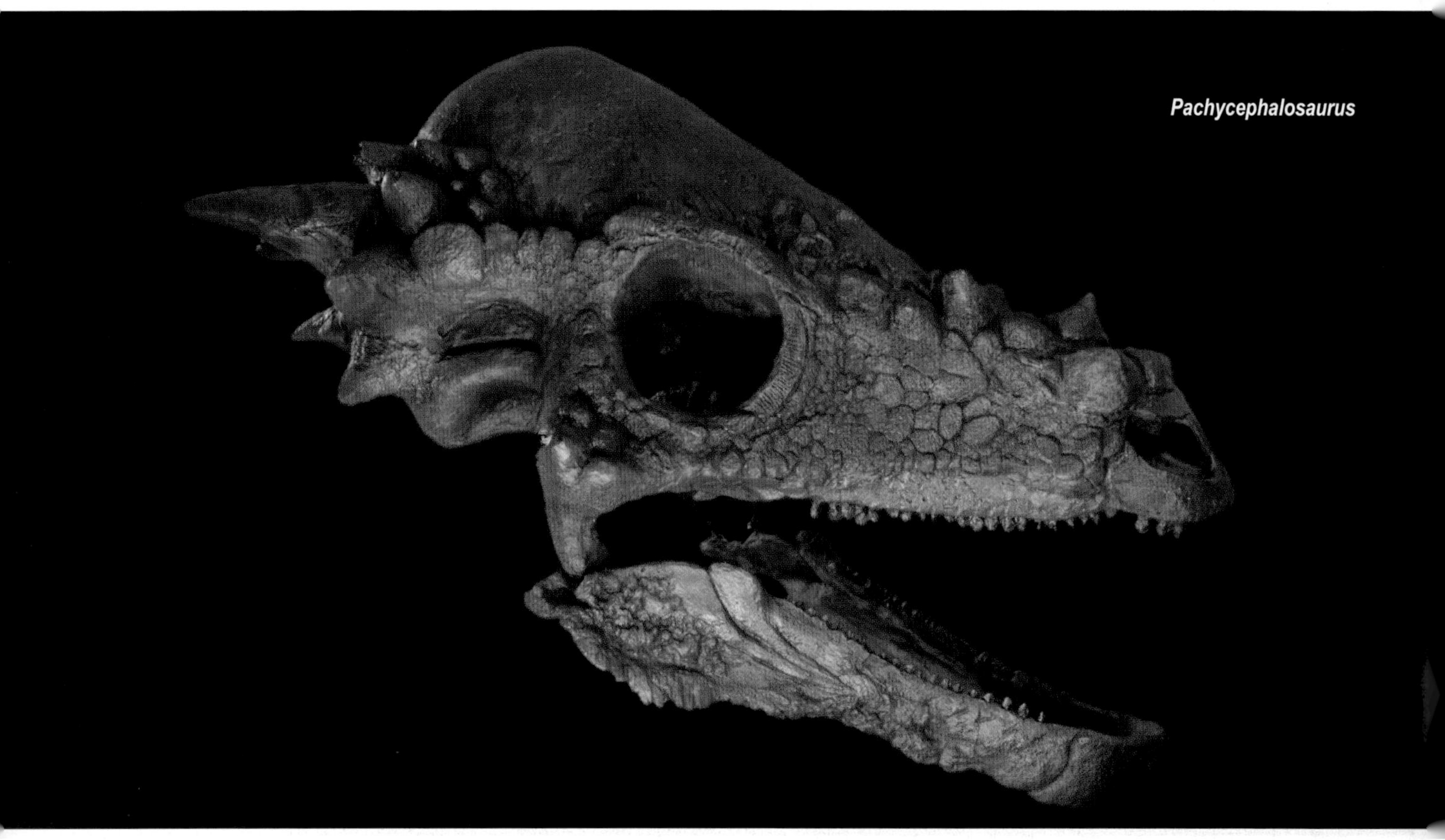

Pachycephalosaurus

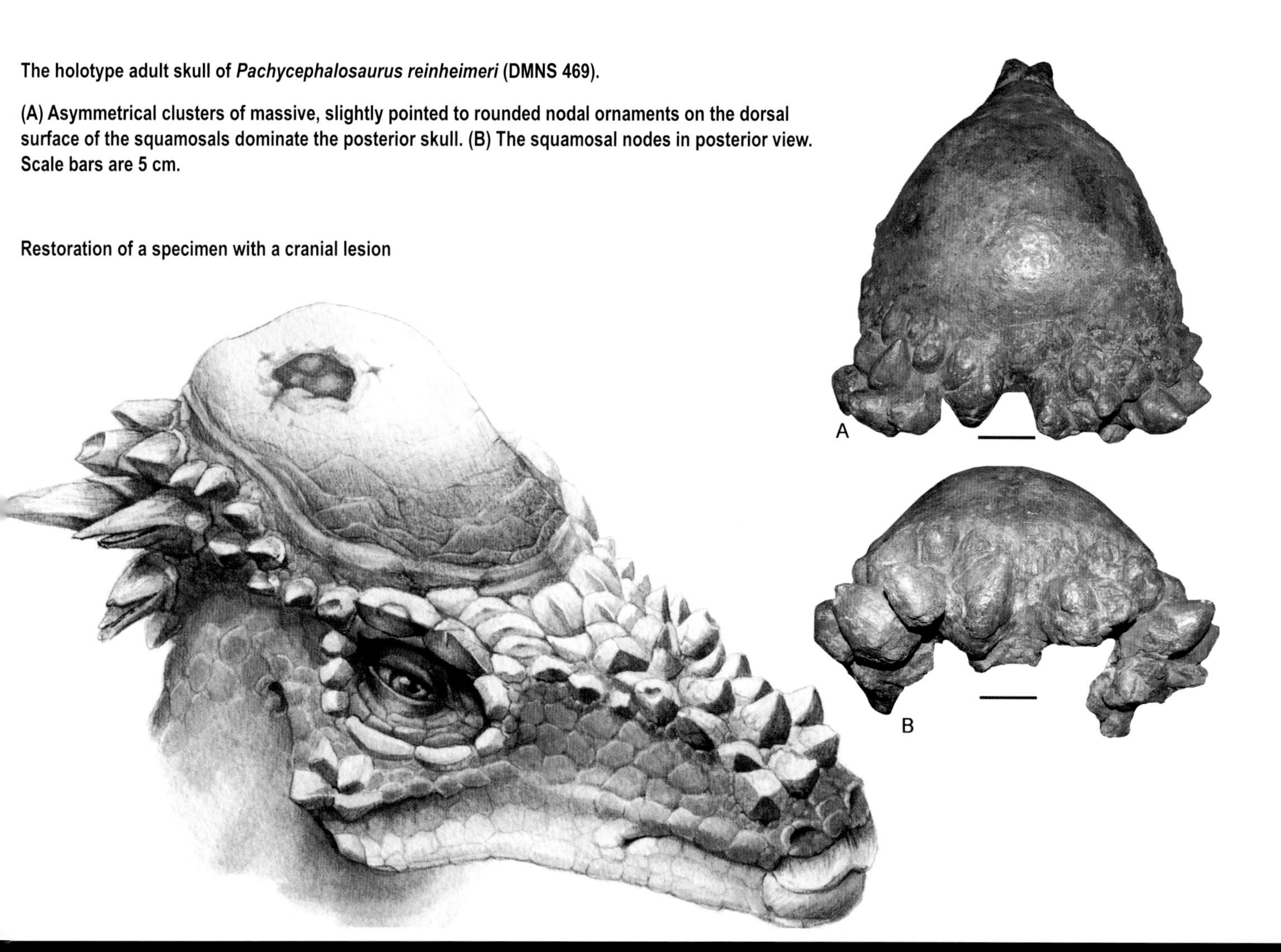

The holotype adult skull of *Pachycephalosaurus reinheimeri* (DMNS 469).

(A) Asymmetrical clusters of massive, slightly pointed to rounded nodal ornaments on the dorsal surface of the squamosals dominate the posterior skull. (B) The squamosal nodes in posterior view. Scale bars are 5 cm.

Restoration of a specimen with a cranial lesion

Dome-headed pachycephalosaurs make a car-bashing appearance in the *Jurassic Park* movies. But did they really use their heads in self-defense? The skull caps of these dinosaurs were up to 20 times thicker than other dinosaurs of the same size. Some of these "domes" were as much as 10 inches thick, but just on the top of the skull. The shape of the skull and the neck seems to indicate that these dinosaurs typically walked with their heads down and the dome pointed forward. Most paleontologists think the thickened skulls were used for head-butting with others of the same kind, as rams and mountain sheep do today. Some think they were used for side-butting, swinging their heads sideways to push other dinosaurs away, even predators. Other paleontologists think the domed skulls were merely for display and/or social status in the herd. God designed these dinosaurs with an unusual thickened skull for possibly all these reasons, and/or just to glorify Him by demonstrating tremendous diversity.

CHAPTER 9: SUBORDER THYREOPHORA: ARMORED AND PLATED

Thyreophora means the "shield bearers." They are distinguished from other dinosaurs because they were designed with one or more rows of bony plates embedded in the skin along the back. All Thyreophorans possessed an ornithischia, or bird-hip pelvis, and nearly all walked exclusively on four legs.

Stegosaurus-like rock carving at Angkor Thom

Scientists divide Thyreophorans into three groups. The first is termed the so-called Primitive Thyreophorans and consists of a few dinosaurs with bony plates, but little else in common with the other two groups, the Stegosaurs and Ankylosaurs. The "primitive" designation is an obvious evolutionary reference, but ancestor/descendant relationships to other Thyreophorans have never been observed. They appear in the Flood sediments, fully formed, and suddenly disappear unchanged, like all dinosaurs.

A second group of Thyreophorans are the Stegosaurs, meaning "plated lizards." These dinosaurs were created with a single or double row of vertical plates or spines arranged along the back. These dinosaurs are common fossils in Jurassic system rocks (late Flood). This group includes the *Stegosaurus*, *Kentrosaurus*, and *Tuojiangosaurus*.

A third group of Thyreophorans were the Ankylosaurs, or "fused lizards." They had small heads and leaf-shaped, nonconnecting teeth similar to the Stegosaurs. They differed because they possessed a nearly complete, turtle-like covering of bony plates across the entire upper body and back. These dinosaurs are further divided into two families, the Nodosauridae and the Ankylosauridae. They are mostly found in Cretaceous system rocks, surviving until late in the Flood.

Saltasaurus illustrates the confusion that can result when trying to classify dinosaurs. As noted in chapter 6, this species had several rows of bony plates along the back, seemingly fitting the definition of a Thyreophoran. However, everything else about the dinosaur fits the definition of a sauropod, including the long neck and tail. The definitions of each group will undoubtedly have to be refined.

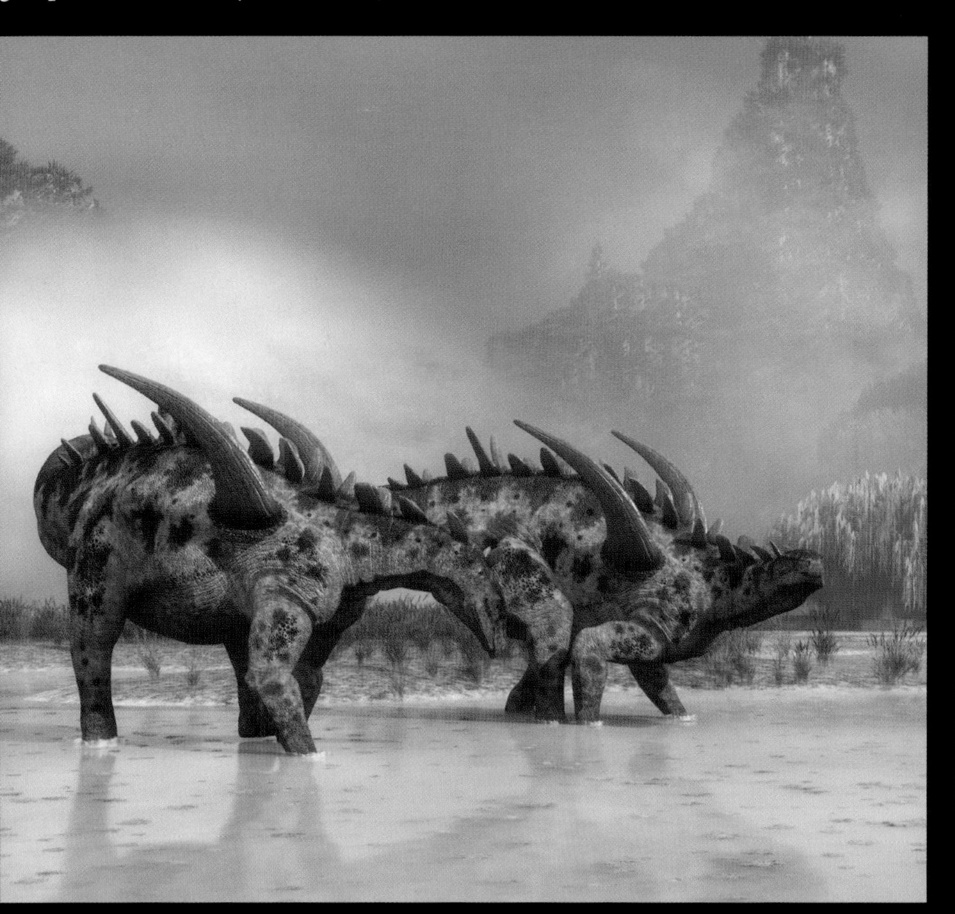

Saltasaurus, an unusual sauropod with bony plates.

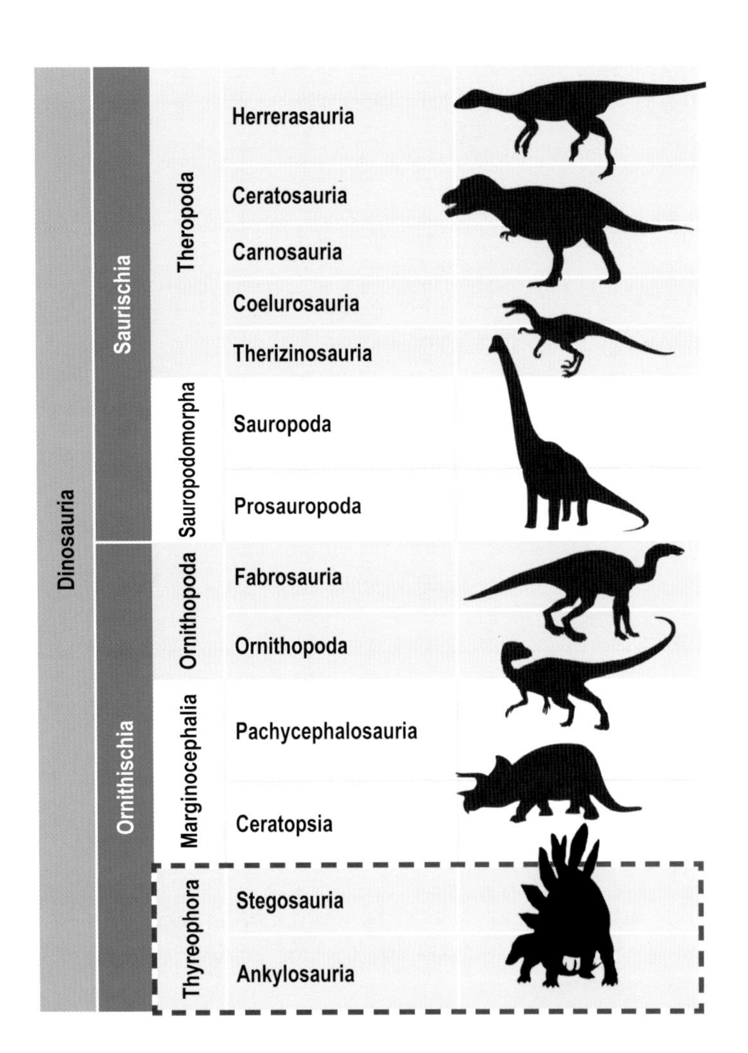

INFRAORDER STEGOSAURIA: *STEGOSAURUS*: THE FAN-SHAPED DINOSAUR

Discoveries of stegosaurs began in England in the 1870s, but were largely composed of isolated bone fragments that were generally unrecognized. It wasn't until more complete specimens were unearthed in the American West by crews funded by O.C. Marsh and E.D. Cope that paleontologists learned how these dinosaurs looked. And they are still not 100 percent sure, even today.

The genus *Stegosaurus* was first named by Marsh from a specimen extracted from the Morrison Formation. *Stegosaurus* has been found in quarries throughout Wyoming and Colorado and has since become the State Fossil of Colorado. It was about 25 to 30 feet in length, designed with an elongated skull, small for its body size, with up to two dozen, leaf-shaped cheek teeth. It did not have front teeth, but instead, had a toothless beak. The eye sockets were rather large and the brain case was very small with a brain estimated at 2.5 to 3.0 ounces. *Stegosaurus* was a low-to-the-ground plant-eating dinosaur, probably similar in eating style to the ankylosaurs. The front legs were designed short and stout with attachment sites for massive shoulder muscles. The hind legs were nearly twice as long as the front legs, giving the animal a fan-shaped appearance from the side.

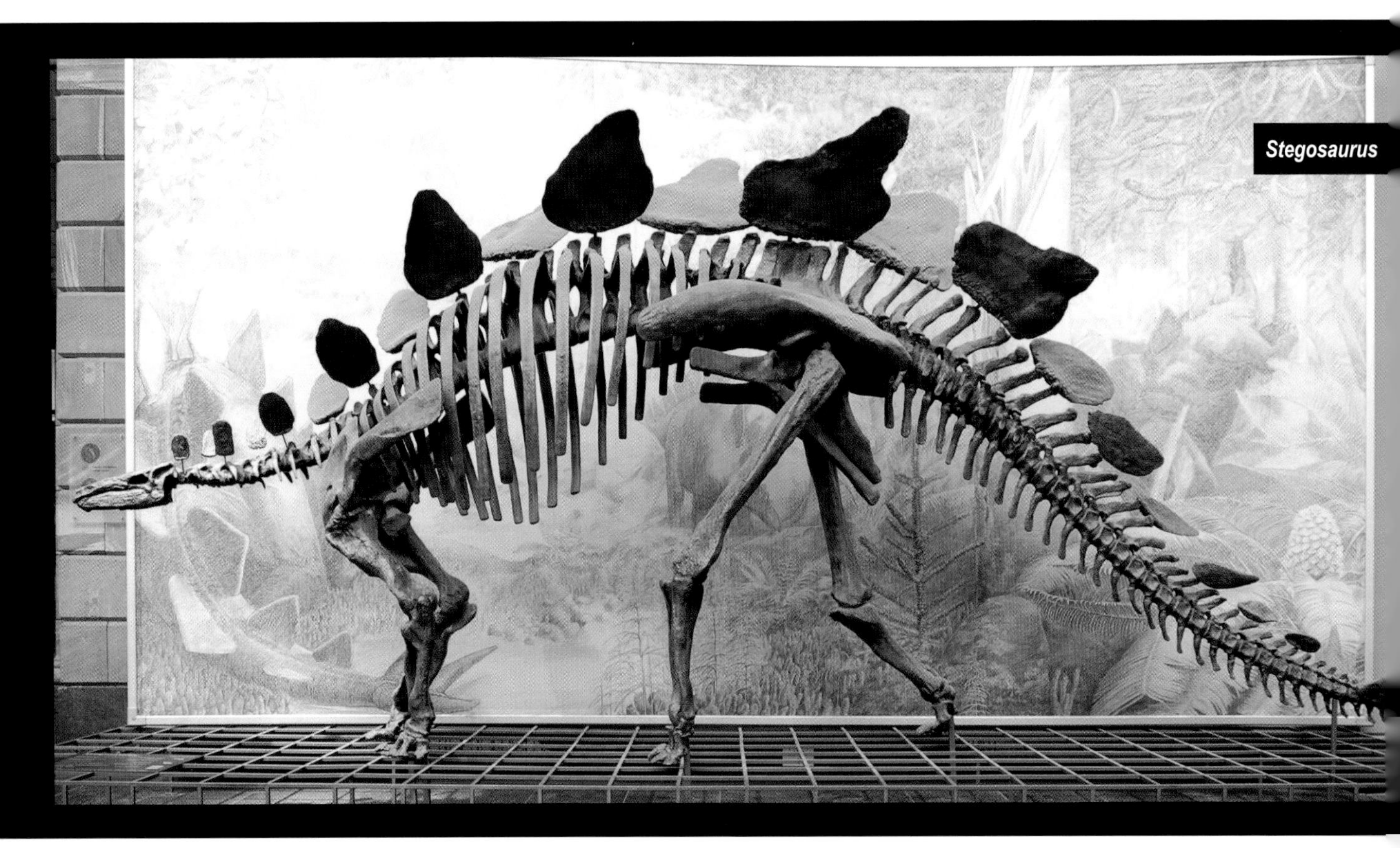

Stegosaurus

The vertical plates protruding along the back must have given this dinosaur quite a large and semicircular appearance in side view. By turning sideways or broadside, *Stegosaurus* would have looked much bigger, deterring predators from attack. *Stegosaurus* also had two pairs of two-foot long spikes at the end of the tail for counterattack.

Over the years, the number of plates and tail spikes has varied in different reconstructions, but most now show four spikes and two rows of plates offset from one another down the back. It had a rather thin body frame when viewed head on. Weight estimates for this dinosaur are around one to two metric tons.

While digging in the Morrison Formation near Thermopolis, Wyoming, I was fortunate to discover a small *Stegosaurus* neck plate. Like all plates, it was highly vascularized, possessing many holes and openings for blood vessels. Most paleontologists believe the plates were arranged in alternating double rows down the back of *Stegosaurus*. These highly porous plates probably didn't serve as a protective barrier, as a bite to the plate likely would have bled profusely. Instead, scientists think the plates served as heat regulators, like a radiator, where blood could flow in and out of the plates to help cool or warm the dinosaur, respectively.

Stegosaurs were designed with a brain slightly smaller than the sauropods whose bodies were so big it made their brain-to-body ratio very small. The brain size and shape was determined by an endocast, creating a replica of the inside of the braincase. The endocast revealed a reptilian-style brain, unlike a bird, with a large olfactory lobe designed for smelling. Their small brain design did not impede their proliferation, as stegosaurs are very common fossils, found worldwide.

It has been suggested that an enlarged cavity in the hind hip region served as a possible second "brain" to help this dinosaur compensate for its small cranial capacity. Strangely, the cavity opening is about 20 times the size of the brain endocast. Most paleontologists now think this hind cavity was for fat and sugar storage as is seen in some birds today, like ostriches.

Stegosaurs were created with five short toes on their front feet and only three toes on their hind feet. Because the hind legs were so much longer, some scientists think they may have been able to rear up on just their hind legs to reach more vegetation. Others believe stegosaurs browsed and ate only close to the ground, remaining on all four legs while eating.

The arrangement pattern of the plates on the back of the genus *Stegosaurus* has been the source of much debate. O.C. Marsh arranged the plates in a single row from head to tail, whereas, most paleontologists think there was some degree of overlap between the plates, and arrange them in an alternating pattern in double rows. The fossils don't really designate which is correct. The 12th-century stegosaur-like carving in Cambodia shows a single row down the back, so that may be the correct pattern. Professor Marsh may have been right all along.

***Stegosaurus* plate found by the author.**

Kentrosaurus and *Tuojiangosaurus* only differed from *Stegosaurus* because they had spines sticking up along the back and not plates. These differences were likely only variations within the created "kind." These two dinosaurs still possessed the distinctive fan-shaped body structure of all stegosaurs, with front legs about half the length of the hind legs.

Kentrosaurus

INFRAORDER ANKYLOSAURIA: *ANKYLOSAURUS*: THE ARMORED TANK

Barnum Brown found the first skeletal remains of *Ankylosaurus* in 1906 in the Hell Creek Formation in eastern Montana. His type specimen consists of a partial skull and a few assorted bones. In 1908, he described and named it the "fused reptile" because of the presence of osteoderms or scutes, bony plates, that grew in the skin of the dinosaur. *Ankylosaurus* is classified with *Stegosaurus* as a Thyreophoran, or shield-bearer. Unlike *Stegosaurus*, this new dinosaur had a solid covering of protective plates, similar to a turtle shell, but composed of bone. Surprisingly, only two other skulls have been found of this dinosaur, along with numerous partial skeletons and scattered osteoderms. All of the fossils have been found in Montana or just to the north in Alberta, Canada. *Ankylosaurus* footprints have been reported near Sucre, Bolivia, but no bones have been found to verify the identification.

Ankylosaurus was the namesake for the family Ankylosauridae. It was designed with bony armor all across the head and back, including horns on the back side of the head, and a large tail club that could serve as a counterattack feature. This club was composed of several osteoderms and supported by ossified tendons and the fusing of the last seven tail vertebrae. Although not very flexible at the end, studies have shown that the tail could still swing from side to side, about 45 degrees in either direction. This club could strike with enough force to break the leg bones of even the largest predator, like *T. rex*. One tail club was found to have chips broken off along the edges, presumably from striking a menacing predator, as evidence of this defensive tactic.

Ankylosaurs, like Nodosaurs, were created with four, fairly short legs. They were the "tanks" of the dinosaurian world and were up to 25 feet long, 6 feet tall, and about 4 feet wide. It is estimated they weighed between 3 and 6 tons. The body armor of bony plates grew in the tough outer skin of the back in rows that fused together into a protective sheath. The skull was also covered in bones, but these seem to have grown outward from the bones of the skull and not in the skin.

***Ankylosaurus* tail club (AMNH 5214), American Museum of Natural History**

***Edmontonia* fossil**

The head was roughly triangular in shape with horns pointing backward from the rear corners of the skull. The teeth were leaf-shaped and show sign of wear from chewing motion both forward and back and side to side.

Ankylosaurs definitely sacrificed escape speed for protection. Their four short legs didn't allow them to run from trouble with any speed and agility, so they likely "hunkered down" when attacked, using the tail club as a counterattack measure.

Ankylosaurus

Nodosaurs were another group of thyreophorans that resembled the ankylosaurs. Nodosaurs differed from Ankylosaurs because they did not possess a bony covering on the top of the skull nor did they have a tail club like Ankylosaurs. These dinosaurs were obligate quadrupeds, walked low to the ground, and were about 18 to 20 feet in length. These two groups were probably the same biblical kind, as they possessed a nearly identical body structure, dimensions, and overall shape.

Over 40 specimens of the species *Euoplocephalus* have been found, making this one of the most common fossil thyreophorans. The obvious body similarities with *Ankylosaurus* make this dinosaur a likely member of the same biblical "kind."

All *Ankylosaurus* bones are found mixed in a rock layer (the Hell Creek Formation) that also contains many ocean-dwelling animals, including 5 species of sharks and 14 species of bony fish. Evolutionists have a difficult time trying to explain these mixed land and marine rock units. Creation scientists easily explain these mixtures as a consequence of the global Flood when tsumani-like waves would be expected to wash marine animals onto the continents and create these deposits mixed with both land and marine animals. In addition, at least three skeletons of nodosaurs have been found in near-shore ocean deposits of shale and chalk. Tsunami-like waves must have washed a few of these dinosaurs out to sea as the Flood waters assuaged.

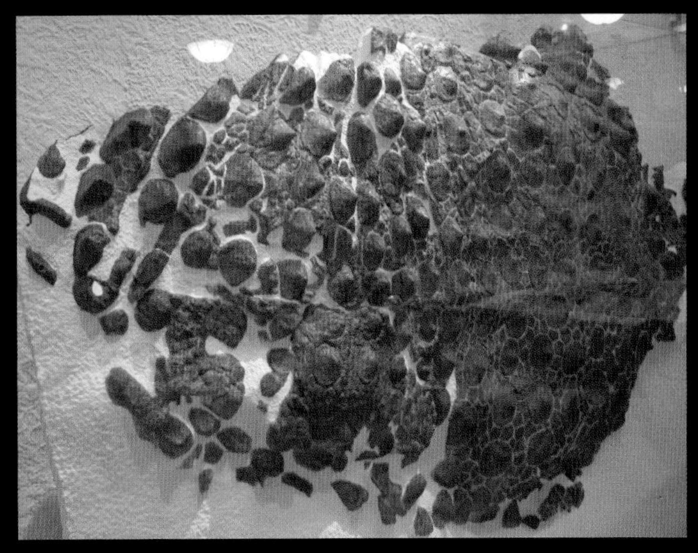

Osteoderm body armor from *Sauropelta*, a nodosaur, American Museum of Natural History.

CHAPTER 10:
DINOSAUR BIOLOGY/ ANATOMY

"Despite nearly two centuries of investigation, a comprehensive understanding of dinosaur biology has proven intractable."[1] This quote from Dr. Greg Erickson summarizes the frustrations inherent in conducting research on extinct animals. Without a living dinosaur to analyze, it becomes difficult to extract physiological data from fossilized material. Studies of dinosaur biology have proven to be even less empirical than study of their fossil bones.

This chapter first reviews the debate over dinosaur metabolism, then covers various topics like growth rates, brain size, and reasons why dinosaurs are not birds, or even bird-related. It also examines how scientists estimate sizes and weights, and reviews the attempts to determine the sex of dinosaurs. We'll even include the math. Finally, the chapter concludes with a discussion of a few more anatomical features such as dinosaur teeth and body covering.

A lot of what is covered in this chapter is hotly contested. Much of it shouldn't be, as you will see, but evolutionists are continually trying to turn dinosaurs into something they are not. We'll expose some of their unscrupulous methods in the process. As scientists, we are supposed to make conclusions and inferences from data we collect or observe, letting the data lead us. Unfortunately, much of secular science is not done the way science is supposed to be done. Evolutionists rarely follow their data; rather, they try and find data to back up their preconceived evolutionary ideas, doing "science" in reverse in the process. This chapter attempts to cut through the evolutionary dogma and instead lets us marvel at the creative design of God's wonderful dinosaurs.

DINOSAURS: DESIGNED WARM-BLOODED OR COLD-BLOODED?

Endothermic (warm-blooded) means that an animal generates its own internal body heat, independent of its surroundings, like mammals and birds do today. Ectothermic (cold-blooded) means that an animal needs an external source of heat, like sunlight, to warm its body, like snakes and lizards do today. Cold-blooded animals cannot maintain a constant body temperature if the external conditions change. These animals need warm temperatures or the sunlight to become active.

Mammals are known as endothermic homeotherms, meaning they have the same, unchanging body temperature throughout. Birds are termed endothermic poikliotherms, meaning they can vary their body temperature by up to 30°F as do turkey vultures.[2] Birds are able to lower their metabolic rate and save energy when necessary, making them more efficient by having to eat less often.

Unfortunately, the proof of whether an animal is endothermic or ectothermic is found only in soft tissue anatomy. Because not enough of this is well preserved in most dinosaur fossils, we still don't know for sure if dinosaurs were warm or cold-blooded. However, the pre-Flood world was likely warmer from pole to pole, making metabolic rate less of an issue before the Flood.

Bone histology involves studying bone tissue by making thin sections across the bone for examination under a microscope. Two types of bone structure have been identified in these studies — lamellar-zonal and fibro-lamellar. Lamellar-zonal bone has a layered appearance with lines of arrested growth (LAGs) or growth lines, and is poorly vascularized, possessing few Haversian canals. LAG bone is most common in modern reptiles and amphibians. The age of these animals can be determined by counting the growth rings or LAGs in the bone, similar to counting tree rings. Fibro-lamellar bone has a fibrous, woven appearance and is highly vascularized, possessing numerous Haversian canals. This type of bone is found in birds, mammals, and most dinosaurs, and contains fewer LAGs.

Highly vascularized bone structure cannot be used as absolute proof of metabolic rate. Fibro-lamellar bone is present in some modern turtles, crocodiles, and lizards, and some dinosaurs contain both fibro-lamellar and lamellar-zonal structures. A study by Tomasz Owerkowicz has shown that the possession of highly vascularized bone merely means the animal was active, and not necessarily endothermic.[3] He found that active lizards had bone microstructure that was highly vascularized, mimicking endothermic animals. Interestingly, Owerkowicz also found that his lizards grew at the same rate as hedgehogs when kept at 95°F, even though their resting metabolic rate was five times slower.

The presence of respiratory turbinates may make a better "proof" of endothermy. Ninety-nine percent of all endothermic animals have turbinates or coils of membrane-covered cartilage or bone in their nasal passages. These structures significantly reduce water and heat loss associated with rapid rates of lung ventilation as needed in endothermic animals.

Although cartilaginous turbinates may not always be preserved as fossils, scientists have found the presence of turbinates is directly correlated with larger nasal cavities in modern endothermic animals. They suspect that larger nasal passages in endotherms serve to accommodate greater lung ventilation rates and provide the room necessary to house the respiratory turbinates. CT-scans on several dinosaur skulls, including two theropods, *Nanotyrannus* and *Ornithomimus*, and one ornithischian, *Hypacrosaurus*, all showed narrow nasal cavities indicative of modern ectothermic animals with little room for respiratory turbinates.[4] The scientists concluded that this is strong evidence for low lung ventilation rates, implying ectothermy or near-ectothermy for these dinosaurs.

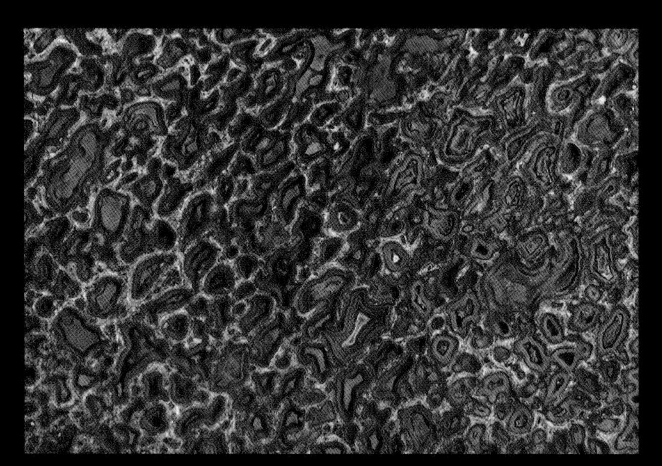

Microscopic close-up of highly vascularized dinosaur bone.

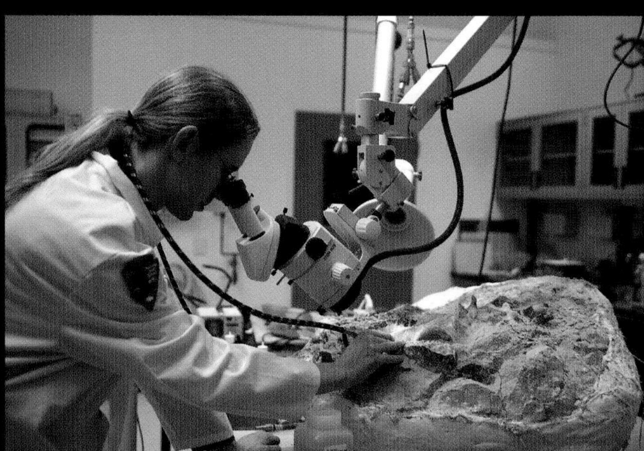

Fossil preparation in the laboratory.

Interestingly, a study on the bird *Archaeopteryx* showed it had distinctive nasal features similar to theropod dinosaurs, implying that *Archaeopteryx* was ectothermic.[5] It's possible that the pre-Flood climate was so mild that endothermy was unnecessary for thermoregulation in these birds and in dinosaurs.

Scientists also found soft-tissue and skeletal support for ectothermy in theropod dinosaurs. Spectacular soft-tissue preservation of the abdominal cavity in the theropod dinosaur *Scipionyx samniticus* indicates a heptic piston, diaphragm-assisted, lung ventilation system similar to modern crocodiles.[6] The attachment style of the intestines and colon also indicates that the avian-style, abdominal air sacs and the flow-through air sac lung were not present in *Scipionyx*. In their study of the theropod, *Sinosauropteryx*, scientists examined the fossilized outline of the abdominal cavity, finding complete thoracic-abdominal separation with a vertically oriented partition that appears coincident with the dome-shaped anterior surface of the liver. This condition is not found in mammals today and is more consistent with the lung system of modern crocodiles. They concluded that diaphragm-assisted lung ventilation (as in crocodiles) was widespread in theropod dinosaurs. They argued that the crocodilian-style lung system and the lack of respiratory turbinates indicate that these dinosaurs routinely maintained ectotherm-like resting metabolic rates. They also found that these theropod dinosaurs may have had the capability of expanding their lung capacity to approach the ventilatory levels of mammals during times of high activity.[7]

A final pitch for endothermy, or warm-bloodedness, in dinosaurs was made by scientists studying teeth enamel.[8] These scientists measured carbon and oxygen stable isotopes "locked" in sauropod teeth, finding that *Brachiosaurus* and *Camarasaurus* had body temperatures near 37°C (98°F). These temperatures were about 5–12°C higher than modern reptiles and within the error range of modern and fossil mammals, but considerably lower than most modern birds (in excess of 40°C). The authors cautioned that their results did not prove endothermy in dinosaurs, as body temperature is a reflection of size, metabolism, and environmental temperature. Creation scientists believe it was much hotter in the pre-Flood world compared to today. These reported temperature results fall within the expected range for a worldwide greenhouse climate, as envisioned for the pre-Flood world.

There is another view that has been proposed that suggests dinosaurs were mesotherms, similar to tuna, lamnid sharks, and the leatherback turtle. Mesotherms can raise their body temperatures by their activity-level, but are unable to internally maintain the temperature like endotherms. In this regard, they are still closer to ectotherms than endotherms. The study relied heavily on bone growth ring (LAGs) data as indicators of dinosaur age for their metabolic rate determinations.[9] Rings in reptile bones, or lines of arrested growth (LAGs), are assumed by evolutionists to be annual, so

The London *Archaeopteryx*, a fossil bird found in 1861 within the Solnhofen Limestone, a fine-grained Flood deposit near Bavaria, Germany.

Artwork of an *Archaeopteryx*. Preserved melanosomes indicate this bird was more likely black like a raven or crow.

Replica of a 131-foot long *Seismosaurus* on display in Thermopolis, Wyoming. Notice author's young daughter for scale.

> Be sober, be vigilant; because your adversary the devil, as a roaring lion, walketh about, seeking whom he may devour
>
> —1 Peter 5:8; KJV

that the age can be determined by counting the rings This may or may not be true for the pre-Flood animals like the dinosaurs we find as fossils. Growth rings, as in trees, and layers in ice cores are unlikely mere annual layers.

In conclusion, there is no compelling reason to make dinosaurs into exclusively warm-blooded animals. I strongly believe dinosaurs were created as cold-blooded reptiles (or at least closer to cold-blooded), like other reptiles alive today. As we shall see later, their cold-blooded metabolism may have played a big part in their eventual post-Flood extinction. The scientific case for endothermy is not strong, and as discussed above, the case for ectothermy is backed by strong scientific evidence.

The unwritten side to this metabolism debate is what is truly fascinating. It is actually a war of worldviews. Mainstream, secular science is trying to make dinosaurs into warm-blooded animals because they are trying to make them the evolutionary ancestors of birds (we will examine this a bit later in this chapter). Birds are warm-blooded, or most are nearly so. For this reason, evolutionary scientists have a compelling reason to push their unfounded and faulty scientific views. It fits their evolutionary worldview. They need dinosaurs to be warm-blooded and feathered if they are going to argue ancestry between the two animal groups. It is difficult to have a warm-blooded descendant from a cold-blooded ancestor.

Unfortunately, many creation scientists have unknowingly become advocates for dinosaur endothermy. They have dismissed the scientific evidence in conflict with this view, and more importantly, failed to notice the deceptive methods of the evolutionary establishment. The trap is set.

DINOSAUR GROWTH RATES

Recent studies of dinosaur growth rates have suggested most species experienced slow growth rates initially, followed by a period of accelerated growth, and finally slowed or reduced growth rates in adulthood.[10] This is called a "sigmoidal" growth curve and is typical of most vertebrate animals. Results show that dinosaurs grew at rates faster than modern reptiles, but slower than birds. Growth rates were determined by counting "growth rings" in bones, similar to counting tree rings. However, in older individuals, medullar expansion causes hollowing of the bones and growth ring loss. Growth rings may be only useful for minimum ages if reliable at all.[11] Therefore, all interpretations concerning the ages of adult dinosaurs and their growth rates must be accepted with some caution. The suggestion that dinosaurs reached a "determinate" size and "plateaued" at a certain age may not be valid either. Growth likely continued, albeit slowly, throughout the adult life of a dinosaur, similar to modern reptiles.

Research has shown that small dinosaurs grew at rates 2 to 7 times slower than similar-sized precocial birds (requiring little parental care), and even slower compared to the faster-growing atricial birds (requiring parental care). Larger dinosaurs grew faster, at rates approaching placental mammals and precocial birds. These data also suggest most dinosaur groups matured rapidly to adult size between 5 and 10 years of age. As expected, some of the smaller dinosaurs matured a bit faster, between 2 and 4 years of age.[12] Tyrannosaurs matured later in life compared to most other dinosaurs, somewhere between 15 and 20 years of age.[13] This suggests that the designed growth pattern of theropod dinosaurs, the purported ancestors of birds, was closer to reptiles instead of comparably sized altricial birds. These data seem to take tyrannosaurs and dromaeosaurs a step further from birds and not closer to them as most paleontologists insist.

The largest dinosaurs were probably well over hundreds of years old, possibly attaining ages like humans in the pre-Flood world. The biggest difference is that the dinosaurs never stopped growing. *Seismosaurus* is one of the longest and likely one of the oldest dinosaurs found to date in the USA. It is estimated to have been 131 feet long. It was found in New Mexico in Jurassic system rocks.

DINOSAURS ARE NOT BIRDS

Thomas Henry Huxley, the famous British evolutionary scientist of the 19th century, was the first person to suggest a dinosaurian origin for birds in publications between 1868 and 1870. He argued that dinosaurs and birds have a similar ankle joint and back foot arrangement with three toes forward and the inner toe backward, a similarly short torso, massively braced hips, a long and mobile neck, and long hind limbs typical of bird anatomy. He also observed that some dinosaurs and pterodactyls had hollow bones, possibly connected to the lungs, as in modern birds. And in a few dinosaurs, he found a backward pubic bone typical of birds. The presence of what is interpreted as a "wishbone" in some theropod dinosaurs was added to this list recently. With cladistics and the construction of cladograms, scientists make these data look even more compelling. However, all of this evidence is circumstantial to say the least.

One of the biggest stumbling blocks to the idea that dinosaurs evolved into birds is the lack of actual fossil support. Birds with real feathers are found in rocks much deeper and buried before the most birdlike dinosaurs. Alleged bird ancestors like the *Velociraptor* and *Deinonychus* are found in rocks dated as Upper Cretaceous system (secular dated at 110 million years old), supposedly 40 million years after the bird *Archaeopteryx* found in Upper Jurassic system rocks (secular dated at 147 million years old). These facts are downplayed by the advocates for the bird/dinosaur relationship who insist that some as yet undiscovered ancestor (called a ghost lineage, because it hasn't been found or seen) must be the common link to both groups. They claim that this unknown ancestor lived prior to *Archaeopteryx*, regardless of the lack of fossil evidence.

Another fact that seems to be overlooked in this controversy is the evidence from dinosaur skin imprints. Literally, scores of dinosaur skin imprints have been discovered in the last 150 years (including reported *T. rex* skin), and none of them show any feathers or feathery projections. The notion that dinosaurs were born with feathers and then somehow lost them as they matured also has no evidence to support it. To the contrary, all embryotic dinosaurs, like the *Saltasaurus* eggs found in Argentina, exhibit scaly, reptilian skin patterns while still in the egg.

In 2000, scientists reported an unusual animal called *Longisquama insignis*, found in Kyrgyzstan, in central Asia. This specimen had clear feathers on its back that were fully developed just like modern birds. The scientists noted how birdlike the animal was with a wishbone similar to that of the bird *Archaeopteryx*. They concluded that *Longisquama* was 70 million years older than *Archaeopteryx*, predating birdlike dinosaurs by 100 million years. They argued that this makes it difficult to imagine that birds could have descended from dinosaurs, since they already existed beforehand. They further doubted that feathers could have evolved twice, once in *Longisquama* and again in dinosaurs, posing even more trouble for evolutionists.[14]

***Protoceratops andrewsi* growth series.**

In spite of what many secularists claim, proto-feathers are not feathers at all, according to some scientists. They found no evidence for the existence of feathers or proto-feathers in any of the published discoveries in China. They determined that the presumed "proto-feathers" were the remains of collagenous fiber "meshworks" that formed feather-looking patterns as a result of decomposition after burial. Their research included analysis of decomposing collagen fibers in modern reptiles, sharks, and dolphins, and supplemental work on several dinosaurs.[15]

More recently, Stephen Czerkas and Alan Feduccia reexamined a purported theropod dinosaur from Mongolia, called *Scansoriopteryx*, concluding unequivocally that it was a bird. They imaged the specimen with advanced 3-D microscopy and high-resolution photography, allowing visualization of features in the wrist bones, feathers, and hind limbs that clearly demonstrated it was a bird and not a dinosaur. As will be discussed below, Stephen Czerkas was earlier involved in the purchase and promotion of the *Archaeoraptor* specimen as a birdlike dinosaur, which later turned out to be a hoax. In 2002, Czerkas, and his wife Sylvia, even self-published a book called *Feathered Dinosaurs and the Origin of Flight,* in which they discussed *Scansoriopteryx*. His most recent article, however, is a complete reversal of his earlier interpretation.[16] It seems Mr. Czerkas has changed his views on at least one specimen after careful examination of the factual data.

Studies of soft-tissue impressions and detailed skeletal analyses also indicate birds and dinosaurs are not related. John Ruben, Terry Jones, and Nicholas Geist examined the lung structure and breathing apparatus of modern birds, crocodiles, and dinosaurs.[17] They found that the meat-eating dinosaurs, including the recently discovered Chinese theropod *Sinosauropteryx*, did not have a bird-type lung-diaphragm, but a crocodilian-type of system. Crocodiles have "septate" lungs, which are a pair of compartmentalized, bellows-type air sacs. They use a liver-diaphragm mechanism to ventilate their lungs and draw in air as the liver is moved forward and backward. Birds, by contrast, use a modified septate system that requires a large sternum and a special hinged rib structure to ventilate their lungs.

The real evidence supports that birds were birds and dinosaurs were dinosaurs from the moment of creation on days 5 and 6. Even the hips of the dinosaurs that are supposedly bird ancestors are "lizard-type" and not "bird-type" hips.

Illustration of a hatchling *Scansoriopteryx*, an extinct bird. Credit Matt Martyniuk.

The Thermopolis specimen of *Archaeopteryx*.

ARCHAEOPTERYX WAS A BIRD

Much of the evidence claimed for a close relationship between birds and dinosaurs comes from comparative anatomy. In the middle of all this controversy is the so-called transitional fossil often cited as the link between birds and dinosaurs, *Archaeopteryx*. As of this writing, there have been 12 specimens found of this amazing creature since 1860. At least 11 of the 12 have been found in a remarkable, fine-grained unit in Germany called the Solnhofen Limestone (Upper Jurassic system).

The alleged relationship between birds and dinosaurs was never clearer (or murkier) than in the history of the *Archaeopteryx* discoveries. Three of the first seven specimens were initially misidentified as either the small dinosaur *Compsognathus* or a pterosaur. Because *Compsognathus* and *Archaeopteryx* are found in the same rocks, showing they lived contemporaneously, it is difficult for evolutionists to argue that *Compsognathus* (the dinosaur) was the ancestor of *Archaeopteryx* (the bird). There are several major skeletal differences between these two animals, notably arm length, and detailed CT-scan analysis of the braincase of *Archaeopteryx* has demonstrated that its brain was birdlike in both shape and relative size. As Pat Shipman pointed out, advocates for the bird/dinosaur hypothesis are left claiming that something yet unknown was the ancestor to *Archaeopteryx*.[18]

Although relatively few specimens of this bird have been found, they have served as the basis for unlimited speculation and controversy. Most evolutionists point to the bony tail, the claws on the wings, and the small teeth in its mouth as "proof" of a dinosaur/bird transition. However, in the minds of most scientists, the well-preserved impressions of designed, asymmetric, flight feathers on several of the specimens, and its overall body structure, have elevated *Archaeopteryx* to status as a true bird. Alan Feduccia, an ornithologist, has pointed out that feathers on *Archaeopteryx* are completely aerodynamic in structure. He claims the evidence doesn't add up somehow and believes that people are merely trying to turn dinosaurs into birds.

In addition, the pubis bone in *Archaeopteryx* resembles modern birds, and was probably used as a muscle attachment site for suprapubic muscles, designed to assist in lung ventilation by moving the tail while roosting. Crocodiles and most theropod dinosaurs, by contrast, utilize their pubis bone as a muscle attachment site for diaphragmatic muscles.

More recently, a structure called a wulst was found in a study of the *Archaeopteryx* brain.[19] This structure, which has only been found in modern birds, is thought to have helped with visualization while flying. Once again, all evidence tells us that *Archaeopteryx* was a true bird.

***Microraptor gui* four-winged bird**

Archaeopteryx also possessed a bird "wishbone," called a furcula, that was found to be robust enough for muscle attachment and capable of flight. Although it doesn't appear to have been a strong flier, its flight might have been enhanced by a more dense, pre-Flood atmosphere. *Archaeopteryx* may have gone extinct rather rapidly, post-Flood, due to changes in climate, and possibly the density of the atmosphere, creating problems for its long-term survival.

Most *Archaeopteryx* specimens show the head arched back in death pose as if gasping for air. Rapid, catastrophic burial during the Flood has commonly produced this pose in many birds and theropod dinosaurs. Finally, the only *Archaeopteryx* on public display in the Western Hemisphere is known as the Thermopolis specimen. It is housed at the Wyoming Dinosaur Center, Thermopolis, Wyoming, USA.

A very strange bird, called *Microraptor gui*, was extracted from Lower Cretaceous system rocks in western Liaoning, China, above the level where *Archaeopteryx* was found. This bird had feathers growing from its hind legs, giving it a four-winged appearance. Scientists are still puzzled as to how this bird used the rear leg feathers while flying. It appears to be another wondrous example of the diversity God gave to His creation.

THE GREAT DINOSAUR-BIRD HOAX

Charles Darwin proposed his version of gradual evolution in 1859. He believed that organisms could mutate into others by natural selection. Nature, he argued, could cause an organism to change due to changes in environmental pressure. When an organism could no longer adapt, it went extinct. Darwin thought that, given enough time, anything could evolve in this manner. Unfortunately, the fossil record never supported his gradual evolution. Darwin believed future scientists would find the necessary transitional fossils to fill in his missing gaps, but here we are, about 150 years later, and there are still no undisputed missing links. Science has had our share of hoaxes, including Piltdown Man and the more recent half-bird, half-dinosaur *Archaeoraptor liaoningensis*, published in *National Geographic Magazine*.[20]

In 1999, the National Geographic Society created a controversy by prematurely announcing the discovery of a half-bird/half-dinosaur they called *Archaeoraptor*.[21] The Society made this announcement of a so-called "feathered" dinosaur prior to thorough peer-review. They did, however, have several famous paleontologists look at the specimen for authenticity. Some of these even coauthored two manuscripts on this new discovery that were submitted to peer-reviewed journals, but were later rejected. It was eventually discovered that the specimen was merely a glued-together composite of a bird and a dinosaur. In other words, it was a complete fake. A Chinese excavator had glued a bird to a dinosaur just to make a bit more money, giving scientists exactly what they wanted in the process.

The National Geographic Society should have waited to announce a "scientific" discovery until after the evidence was scrutinized for scientific integrity. This specimen was not even collected in the field, as is proper, leaving no way to verify its true authenticity. Stephen Czerkas had the Dinosaur Museum in Blanding, Utah, purchase the unusual Chinese specimen from an anonymous dealer at the Tucson Gem Show in 1999 for $80,000.

Part of the reason prominent scientists fell for the *Archaeoraptor* hoax was their belief in their own hype, regardless of the poor science involved. They believed so strongly that birds evolved from dinosaurs that all evidence to the contrary was ignored or brushed aside as unimportant.

> For the time will come when they will not endure sound doctrine, but according to their own desires, because they have itching ears, they will heap up for themselves teachers; and they will turn their ears away from the truth, and be turned aside to fables
>
> —2 Tim. 4:3–4

Image of *Archaeoraptor* from *National Geographic* magazine (Nov. 1999). The specimen turned out to be an extinct bird and a dinosaur merely glued together.

Jonathan Wells later labeled the specimen the "Piltdown Bird" in memory of the infamous Piltdown Man hoax.[22] To their credit, the National Geographic Society did publish a short retraction the following year.[23]

Because of the nagging lack of transitional fossils, Stephen J. Gould and Niles Eldredge proposed a modification of Darwin's original theory that tried to explain away missing links, called punctuated equilibrium. Their hypothesis didn't really provide actual rock and fossil evidence; rather, it was merely an acknowledgement that the fossil record shows repetitive episodes of no change (stasis) followed by periods of sudden change. This is precisely what creationists have pointed to as conclusive evidence of a catastrophic Flood, with sudden fossil changes occurring in rapid succession as sediments accumulated. It all depends on your starting assumptions, your worldview, to explain the same results.

The 2003 discovery of the fossil fish *Tiktaalik* was claimed by evolutionary scientists as the "missing link" between fish and amphibians, and the first animal that could walk as a tetrapod (four-legged). However, the weak backbone in the pelvic region and the presence of fins, not fingers or claws, would have made it unable to stand or walk. Finally, the subsequent discovery of tetrapod footprints in rocks that predate *Tiktaalik* suggest it was just another odd-looking pre-Flood fish, and not a missing link, and not the "first" tetrapod. The evolutionists need to go back to the "drawing board," again.

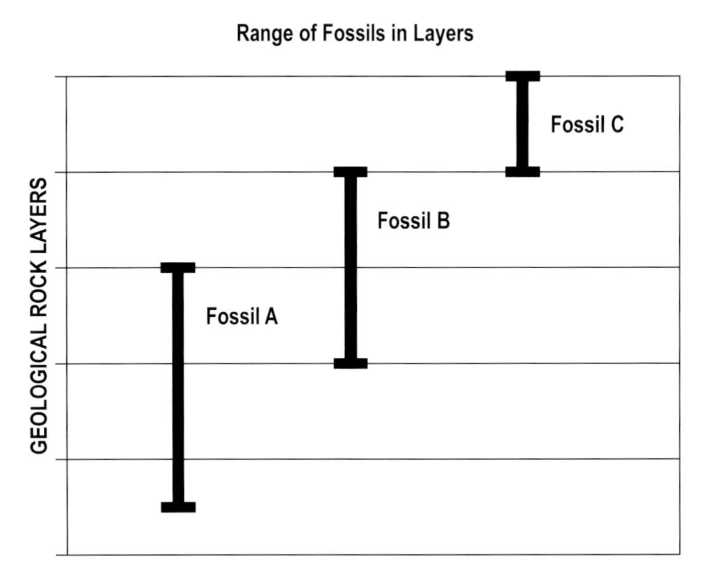

Fossils appear in order of their burial in the Flood. They appear suddenly, remain the same, and disappear just as suddenly.

Tiktaalik

BRAINS AND BRAWN: WHAT WERE DINOSAUR BRAINS REALLY LIKE?

A lot of misinformation concerning dinosaurs is shown in the movies and in popular media and even slips into documentaries or "scientific" media, pushing the evolutionary agenda. Evolutionists are continually insisting that dinosaurs evolved into birds, despite the strong evidence against it. One of the misconceptions that is portrayed concerns the brains of large theropod dinosaurs like *T. rex*. However, CT scans of the skulls of *T. rex* have given scientists additional details of the brain cavity, demonstrating its large olfactory lobe (for smell) and its overall shape that is similar to modern alligators and not birds.

Birds have a completely different shaped brain, with a larger section of the brain for processing data, not just for sensing data. Birds have to do more than sense a food source, they have to be able to discern one food source from another. Alligators just smell something and snap at it, without thinking about whether or not it is a good food source. Not only is a bird brain shaped differently, but pound for pound, the typical brain of a bird is much larger, by nearly an order of magnitude (or ten times), compared to a typical reptile brain.

To examine this brain situation more scientifically, I obtained a full-scale model of a *T. rex* brain cast, called an endocast. The brain cast was about 8 inches in length. I estimated the encephalic volume (brain size) of the adult *T. rex* by dunking the brain cast in water. The volume of water displaced, and therefore the brain volume, was found to be about 158 cc (cubic centimeters). I then compared this value to the size expected for a typical reptile, scaled up to the size of a *T. rex*. My results showed that the brain volume fell very close to the expected value for a reptile, demonstrating that the *T. rex* was not birdlike in brain size either. For comparison, the average human adult male brain has a volume of around 1273 cc, giving *T. rex* a brain nearly an order of magnitude smaller than our own human brain.

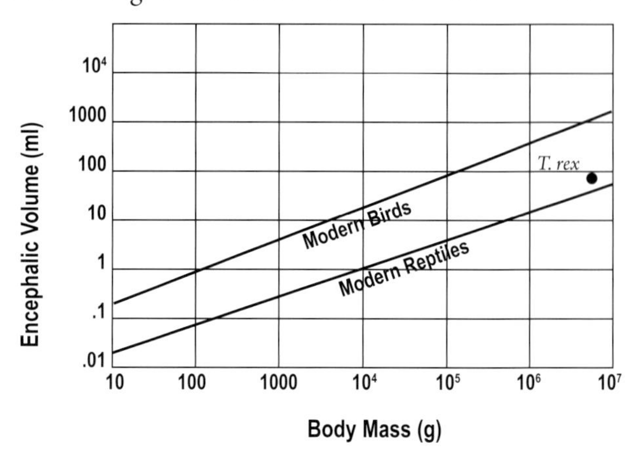

Log-log plot of encephalic (brain) volume v. adult body mass showing straight-line relationship of modern birds and reptiles. Birds have an order of magnitude greater brain-to-body ratio compared to reptiles. An adult *T. rex* plots much closer to the reptile line.

Brain endocasts may not be able to give us a true measure of intelligence, but we do see that dinosaurs had different brain sizes relative to their body size. Stegosaurs and sauropods had the smallest brain-to-body ratios, whereas theropods like *T. rex* had much larger brain-to-body ratios. Apparently, the stegosaurs and sauropods were designed with smaller brains because that is all they needed to locate their food, function, and reproduce. Theropods had to either see or smell to hunt their food sources, so they were designed with larger sensory powers that resulted in a larger brain.

Unfortunately, what you see in movies and often in print is a lot of misinformation to make dinosaurs something they are not. As we shall see in the next section, they are not birds. Their brains were shaped and sized more like typical reptiles. All available data on brain size and shape indicate that their intelligence, thought processing, and senses were probably similar to alligators. Dinosaurs may have been able to smell really well, but they could not open doors.

God created all dinosaurs on day 6 of creation week (Gen. 1:25). He designed dinosaur brains perfectly suited for their lifestyle. Their brains were in proportion to their reptilian bodies and only as large as they had to be to function as He designed. Only our own species was created in the image of God (Gen. 1:26), and given dominion over all other created animals.

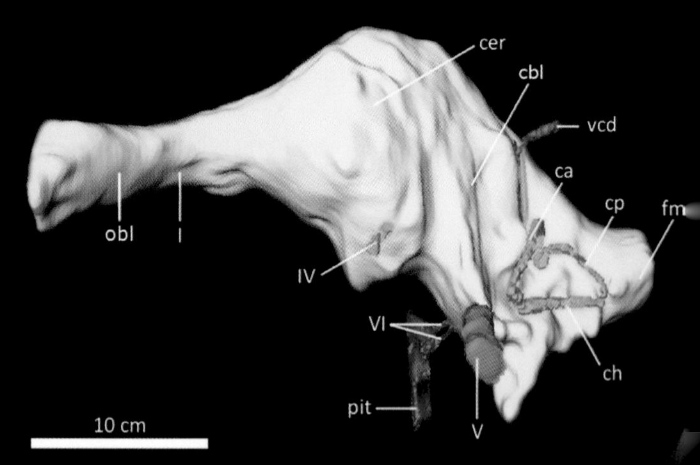

Digital endocranial endocast of the braincase of *Acrocanthosaurus atokensis* (NCSM 14345) in left lateral view.

ca, anterior semicircular canal; cbl, cerebellum; cer, cerebrum; ch, horizontal semicircular canal; cp, posterior semicircular canal; fm, foramen magnum; I, olfactory nerve; IV, trochlear nerve; obl, olfactory bulbs; pit, pituitary; V, trigeminal nerve; vcd, vena capita dorsalis; VI, abducens nerve.

MEDULLARY BONE: THE SECRET TO DINOSAUR SEX DETERMINATION

Dr. Mary Schweitzer was the first to report the discovery of dinosaurian medullary bone.[24] This type of bone forms in leg bones of some female birds before ovulation as a calcium reservoir for eggshell production. Unfortunately, medullary bone is quickly reabsorbed after ovulation. Therefore, a specimen that possesses it must be fossilized at precisely the right moment of pregnancy.

At least three dinosaurs have been found to possess medullary bone, and were therefore pregnant at the time of death — *Tyrannosaurus rex*, *Allosaurus fragilis* (both theropods), and *Tenontosaurus tilletti* (a duckbill). By counting bone growth rings, scientists were able to estimate that reproductive maturity for these dinosaurs began about the time their own growth began to slow, or just after their growth spurt ended. *Tyrannosaurus* was found to be sexually mature at 18 years, *Allosaurus* was found to be sexually mature at 10 years, and *Tenontosaurus* was found to be sexually mature at 8 years of age.[25] This implies the reproductive strategy of dinosaurs was surprisingly similar to living reptiles. In both, the onset of sexual maturity occurs before the growth curve completely flattens.

There is no other way at present, other than the presence of medullary bone, to determine the sex of a dinosaur. The hip bones are not structured differently in the sexes like they are in many mammal species. Previously (chapter 5), I discussed Peter Larson's suggestion that *T. rex* can be divided into two morphotypes he called gracile and robust. He determined that the vertebrae behind the pelvis differed from type to type. By comparing the differences in vertebrae behind the pelvis to crocodiles, he inferred the larger, robust *T. rex* was a female. However, it was also found that these perceived differences in crocodile vertebrae did not always hold true to the sex.

It is interesting to note that the pregnant *Tyrannosaurus rex* studied by Dr. Schweitzer was a relatively small specimen with a femur length of about 42 inches.[26] This suggests that the gracile *T. rex* morphotype was female. It further implies that the larger, more robust *T. rex* skeletons, like the famous "Sue" specimen in Chicago, were most likely males, opposite to popular theory and Peter Larson's speculation.

Sexual dimorphism, or physiological differences between males and females of the same species, is thought to be common among the dinosaur species. Unfortunately, in most cases scientists don't know which one is the male and which one is the female. Unrecognized, this factor may have further complicated the naming of species by calling the same dinosaur two different names due to their physical differences.

God created males and females of every kind. And males and females of each kind of dinosaur were brought to the ark by God to preserve each kind.

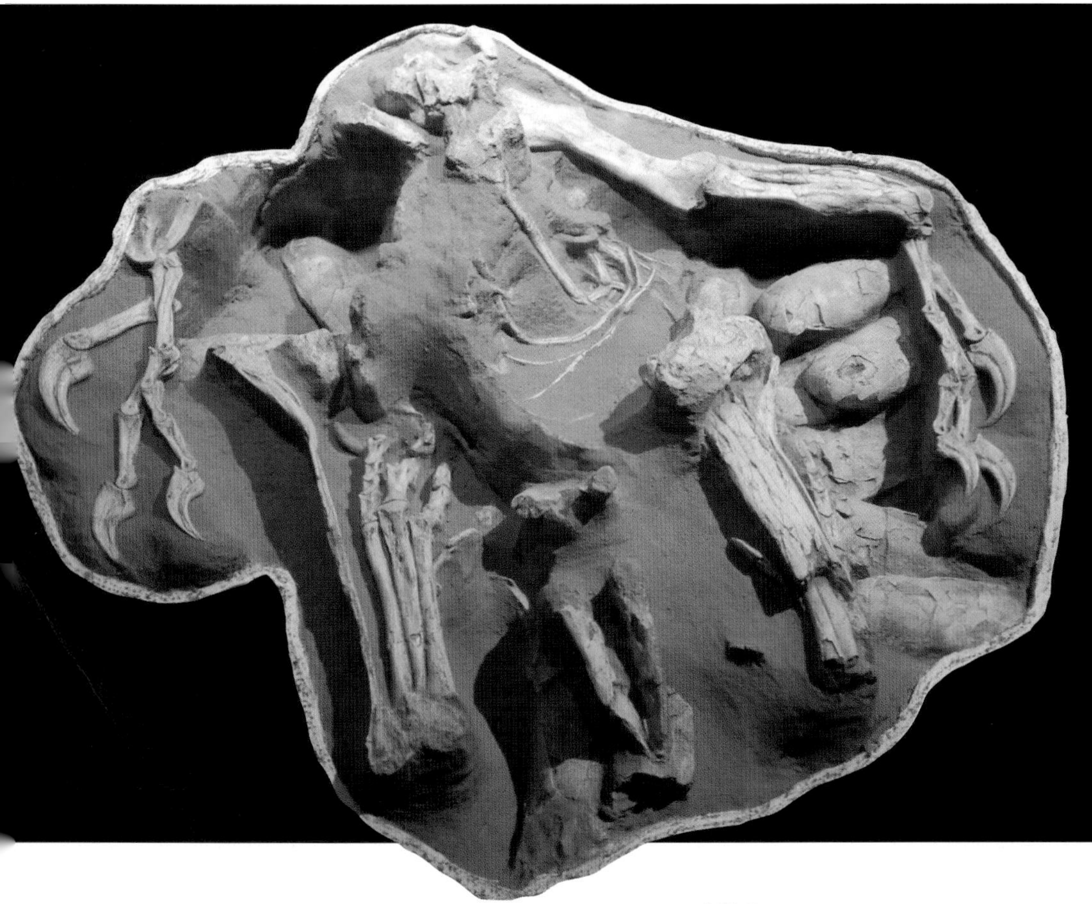

Photograph of an *Oviraptor* found buried in Flood sediments while sitting on a nest of eggs. The dinosaur may have just finished laying the eggs before it succumbed to rapid burial. Specimen IGM 100/979 found in Mongolia by the American Museum of Natural History, New York.

HOW HEAVY IS THAT DINOSAUR?

Determining the exact weight of a dinosaur is something we'll never be able to do without a living specimen. Scientists can only make estimates using skeletal data. For this reason, weight estimates for some dinosaurs have changed dramatically over the years. The earliest reconstructions made dinosaurs out to be fat, sluggish, lethargic-looking lizards. Most of the first reconstructions of upright dinosaurs were in the classic, kangaroo-like pose, with the tail dragging on the ground behind. More recent portrayals have made dinosaurs leaner and more agile-moving creatures. Detailed computer analyses, footprint data, and movies such as *Jurassic Park* have made these slimmer reconstructions mainstream.

There are two generally accepted ways to estimate dinosaur weights: (1) use of leg bone circumferences, and (2) use of fleshed-out scale models. Most paleontologists prefer to use scale models, but there are distinct advantages and disadvantages to both methods. The beauty of the circumference method is in its simplicity. All a paleontologist needs is a couple of leg bones, which are often well-preserved, in order to make a reasonable estimate of body weight. The scale model method requires a more complete skeleton to begin with, but may give better results.

Scientists have worked out a method to estimate the weight of a quadruped (4-legged) based on the circumferences of the humerus and femur bones of the upper legs, measured at the midpoint where the bones are the thinnest. They summed the two circumferences and plotted the value against body mass on a logarithmic scale and found a straight-line relationship and corresponding best-fit equation for us to use.

The resulting **quadruped equation** is:

Body mass in kg = (0.000084) x (humerus plus femur circumference in mm)$^{2.73}$

Note that the total circumference is not squared, but is raised to the 2.73 power. This equation can be used to estimate the weight of any quadruped dinosaur such as *Diplodocus* or *Brontosaurus*. However, it is based solely on modern mammals, and extrapolations to dinosaurs may lead to considerable error.

These same scientists have also provided an equation for bipeds (2-legged) animals. In this case, they used only the femur circumference (hind leg).

The resulting **biped equation** is:

Body mass in kg = (0.00016) x (femur circumference in mm)$^{2.73}$

Because *Tyrannosaurus rex* is obligate bipedal, we use this equation for its weight estimate. All that is needed is a femur. Like the last equation, this equation is based on modern animal data, and caution must be exercised in any estimate of dinosaur weight.

One problem that is not addressed by the two equations is the "in-between" dinosaurs that walk on four legs, but run on two. Some of the hadrosaurs (duck-bills) were neither obligate quadrupeds nor obligate bipeds. Weight estimates using the circumference method for these types of dinosaurs will undoubtedly cause greater error. My previous estimate of "Eddie" the hadrosaur may be off a bit because of this factor (chapter 1).

The scale model method starts with a solid, waterproof, plastic scale model. The method involves determining the displacement volume of the model in water and scaling up the three dimensions of the model to the full, life scale of the dinosaur. Refer to the appendix for the details.

ARGENTINE DINOSAURS: THE BIGGEST OF ALL

Argentina has produced some of the largest and most noteworthy dinosaurs of the past few decades, and still claims two of the earliest dinosaurs (buried the deepest in the Flood sediments). *Eoraptor* and *Herrerasaurus* were found in rocks of the Upper Triassic system. Both of these reptiles were bipedal and likely theropod dinosaurs. *Eoraptor* was about 3 feet long and *Herrerasaurus* was about 10 to 20 feet in length. The "oldest" dinosaur, the one buried the earliest and deepest, now seems to belong to a Tanzanian dinosaur found in Middle Triassic system strata.

One of the largest collections of dinosaur eggs and nesting sites in the world was found in Cretaceous system rocks in Neuquen, Argentina. The site covers about a square mile in area and is littered with thousands of eggs, eggshell fragments, and nests, including numerous ones with embryos of a titanosaur sauropod (*Saltasaurus*?). Well-preserved imprints of scaly, embryotic skin were also found in the eggs. The embryos were about 15 inches long, found in round eggs about 6 inches in diameter. Although no adult bones were found at the site, it is believed that the adults grew to about 40 feet long. Scientists interpret this find as the site of a catastrophe where the embryos in the eggs were drowned by flooding and encased by silty flood sediment. Creation scientists couldn't agree more, except to add that burial occurred during the great Flood just thousands of years ago.

Huge vertebrae and a few other bones from a dinosaur called *Argentinosaurus* have been excavated from northern Patagonia, Argentina, a site that seems to be continually in the news. The bone dimensions of this titanosaur sauropod indicate that it may have weighed close to 70 metric tons. This would make this the heaviest dinosaur of all yet discovered (although a newer discovery of a similar titanosaur in the same region may push the heaviest dinosaur to 77 metric tons). Another partial skeleton of an even longer sauropod was discovered in Patagonia, Argentina, that may have been as long as 167 feet long.

The "heaviest" dinosaur with the most reliable size and weight, as it was found nearly 70 percent complete, is another Argentine dinosaur from the same region called *Dreadnoughtus schrani*. This huge titanosaur was estimated at 59.3 metric tons and 85 feet long. The paleontologists who analyzed this individual determined that this dinosaur was still growing at the time of death, as its shoulder bones had not yet fused, as observed in more mature adult sauropods.

Undoubtedly, scientists will continue to find bigger dinosaurs in the coming decades. Because dinosaurs, like all reptiles, grew throughout their lifetime, these gigantic animals were probably the oldest of their kind to be trapped and buried at the time of the Flood, and may have been hundreds of years old.

Argentinosaurus huinculensis

Argentinosaurus skeleton

Giganotosaurus is the second largest theropod, and assumed meat eater, found to date. Bones from four or five of these dinosaurs were found together in Neuquen, Argentina, from what are called Middle Cretaceous system rocks (Late Flood). *Giganotosaurus* was a few feet longer than the biggest *T. rex*, and it is estimated to have weighed 6 to 8 metric tons. *Giganotosaurus* had an overall body shape that closely resembled a *T. rex* except it had thinner, shark-like teeth and three claws on its small upper arms. The discovery at the site of two adult specimens with a few smaller members of the same species has paleontologists thinking an entire family unit was swept away by flood waters and buried together, again, providing more evidence for the great Flood.

Another Argentine dinosaur that was previously discussed, the 40-foot long *Saltasaurus*, was an unusual sauropod that had bony plates in rows along its back. The adults weighed in at about 7 metric tons and were found in Cretaceous system rocks.

Argentina also produced one of the largest prosauropod dinosaurs, including the massively built 35-foot long *Riojasaurus*. This dinosaur was excavated from rocks of the Triassic system. Numerous other sauropod dinosaurs are also found in Argentina. These include *Patagosaurus* from Middle Jurassic system rocks and *Rayososaurus* and *Amargosaurus* from Cretaceous system rocks (all from later Flood sediments).

Partial femur from *Argentinosaurus*. This dinosaur may have weighed as much as 70 metric tons and been over 30 feet tall. The complete femur would have been close to 8 feet long.

DINOSAUR TEETH

Nearly all dinosaurs were designed to grow new teeth throughout their lifetime. Some have estimated that an individual tooth might have lasted only a few years before falling/breaking off. For this reason, many more tips of shed teeth are found compared to those with roots. I found two shed *Allosaurus* teeth in Wyoming. Both were black but looked fresh and still sharp. Neither one had the root attached. Teeth are also more durable than typical bone because the enamel (made of calcium phosphate or apatite) is made of a tight, compact structure.

Paleontologists estimate the type of food an animal ate by examining its teeth and jaw structure. Some teeth were leaflike, some were peglike, and others were conical in shape. Most theropods are believed to have been meat eaters, as they often had conical, serrated teeth like "steak knives." Some theropods, like *Oviraptor*, had no teeth, however, just beak-like jaws. Others, like the *T. rex*, had large, spike-shaped teeth, whereas other theropods had razor-sharp and thin teeth, like *Velociraptor* and *Allosaurus*. Many teeth were curved backward to hold prey from escaping prior to swallowing.

Teeth of other fossil reptiles, like crocodiles and alligators, were designed much differently compared to dinosaurs. Extinct reptiles have fairly straight, hollow teeth without serrations. Many had grooves or striations running the length of their teeth. Most reptilian teeth were used for crushing the prey, not grinding. Crocodiles and alligators replace their teeth throughout their lifetimes, like dinosaurs.

Some sauropod (long-necked) dinosaurs were designed with only peg-like teeth in the front of the mouth, and no molars to chew with. Dinosaurs like *Brontosaurus* and *Diplodocus* presumably stripped vegetation from branches and swallowed it whole. These dinosaurs are commonly associated with gizzard stones for use in digesting the swallowed food.

Camarasaurus and other similar dinosaurs were created with leaf or spoon-shaped teeth that extended to the back of the mouth. These teeth allowed the dinosaurs to take a better "bite" and to do a bit more chewing compared to the peg-shaped teeth.

Tyrannosaurus rex had the largest teeth and root design of any theropod dinosaur found to date. They were not just long, up to 6 to 12 inches, but were thick, more like spikes. Most theropods, by contrast, had thinner knife or sword-shaped teeth. *T. rex* probably ate differently compared to most predators, chomping down and crushing bone and not slashing.

Giganotosaurus, found in South America in the 1990s, was even bigger than *T. rex*, but it had smaller, thinner, razor-like teeth more typical of most theropod dinosaurs. These dinosaurs likely slashed their prey like great white sharks.

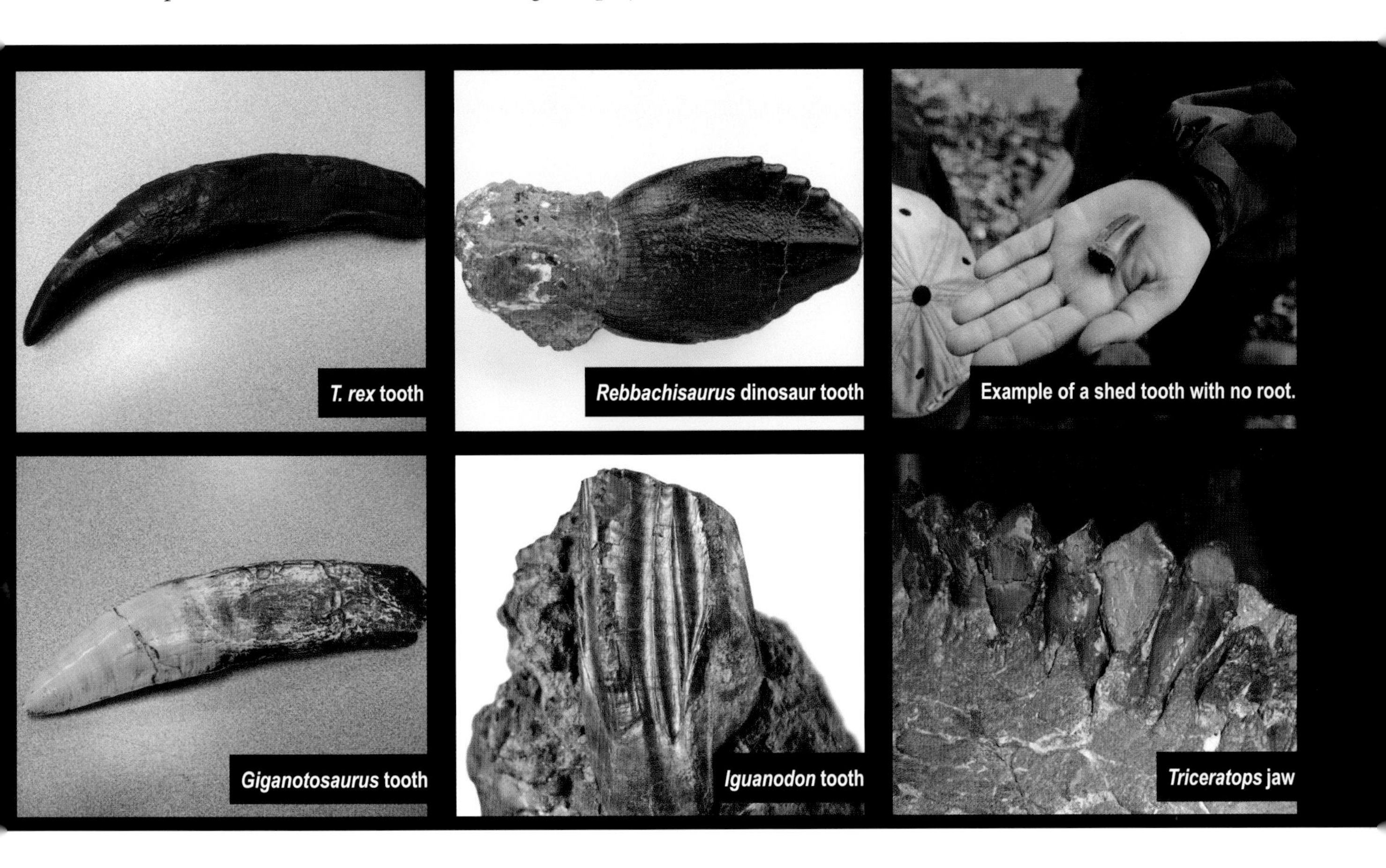

T. rex tooth

Rebbachisaurus dinosaur tooth

Example of a shed tooth with no root.

Giganotosaurus tooth

Iguanodon tooth

Triceratops jaw

Many of the dinosaurs, like the hadrosaurs (duck-billed) and the ceratopsians (horned-dinosaurs), had back jaws (cheek teeth) filled with large brackets of chewing and grinding teeth capable of eating very tough vegetation, like bamboo and rushes. These teeth were self-sharpening, as they were softer on one side and would wear unevenly as they bit down and ground against each other, acting like scissors when biting.

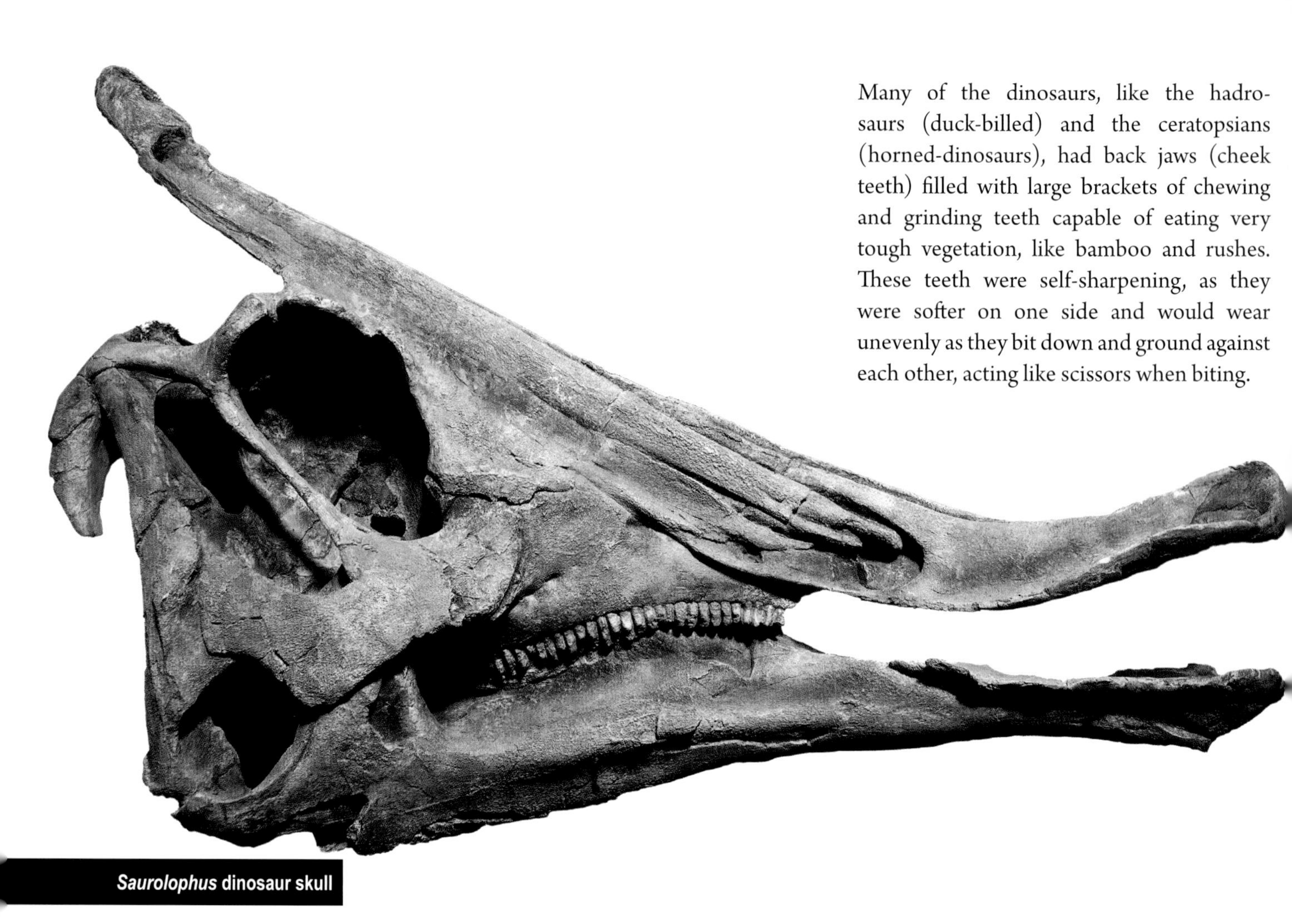

***Saurolophus* dinosaur skull**

Paleontologist Rodolfo Coria dusts teeth of *Giganotosaurus*, a theropod dinosaur slightly longer than *T. rex*. The first specimen of this dinosaur was found near El Chocon, Argentina, in 1993. This photograph was taken in 1995.

DINOSAUR SKIN AND SCALES

Numerous fossilized "mummy" dinosaurs have been found that have preserved spectacular skin impressions. Many of these are from hadrosaur dinosaurs (duck-bills). All show a scaly pattern of lizard-like skin. One of the more famous of these has been dubbed "Dakota," as it was found in North Dakota. The pattern of scales on Dakota shows banding between the larger and smaller scales. Scientists couldn't derive color, but noticed a striped pattern to the skin when looking at it in monochrome light.

Dakota also surprised paleontologists because the well-preserved rump and back end of the dinosaur appeared to be about 25 percent larger than previously thought. Scientists speculated that this meant greater muscle mass, and likely greater acceleration when evading an attack. In truth, scientists have little knowledge of how fat or thin most dinosaurs were, or if they had flaps of skin like *Dilophosaurus* exhibits in the movies.

The best information to date comes from the best-preserved fossils, like Dakota.

A similar hadrosaur "mummy" named Leonardo, was unearthed in Montana. It also showed magnificent preservation of skin, muscles, and internal tissue, including the stomach. Leonardo had polygonal, scaly skin like Dakota. Both of these specimens have been called "breath-taking" and "once-in-a-lifetime finds" for their spectacular state of preservation. In the past, paleontologists thought these specimens were exposed at the surface for weeks and allowed to dry out before burial in sediment. Now, both are thought to have been rapidly encased in wet sediment, thus creating an environment suitable for such detailed preservation. Creation scientists have argued for this position all along, calling on rapid burial in Flood sediments to preserve dinosaur fossils of all kinds.

Other mummified specimens have been found over the past 100 years in the American West and Canada. None of these well-preserved skin fossils have shown any hint of feathers or even proto-feathers, as advocated by many secular paleontologists and even some creation paleontologists. In fact, a 2017 study showed conclusively that all tyrannosaurs, including *Tyrannosaurus rex*, had scaly, reptilian-like skin and no feathers whatsoever.[27]

Well-preserved fossil dinosaur skin from a plant-eating *Psittacosaurus* was studied and found to contain 40 separate layers.[28] Each layer was designed with collagenous fibers arranged in opposite directions, layer to layer, giving strength to the skin similar to the cross-grain strength ability of plywood. This design provided a tough outer covering for these dinosaurs to serve as protection from predators.

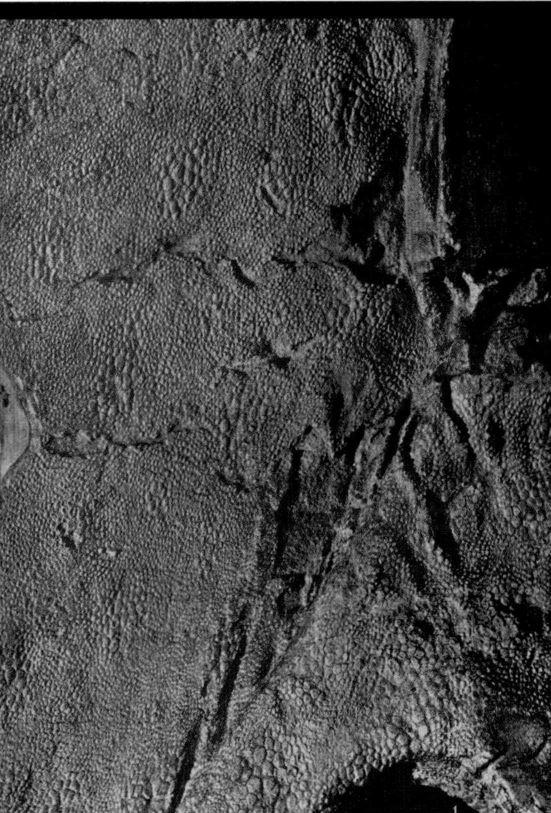

Skin impressions of a duck-billed dinosaur photographed in 1912; now at the American Museum of Natural History.

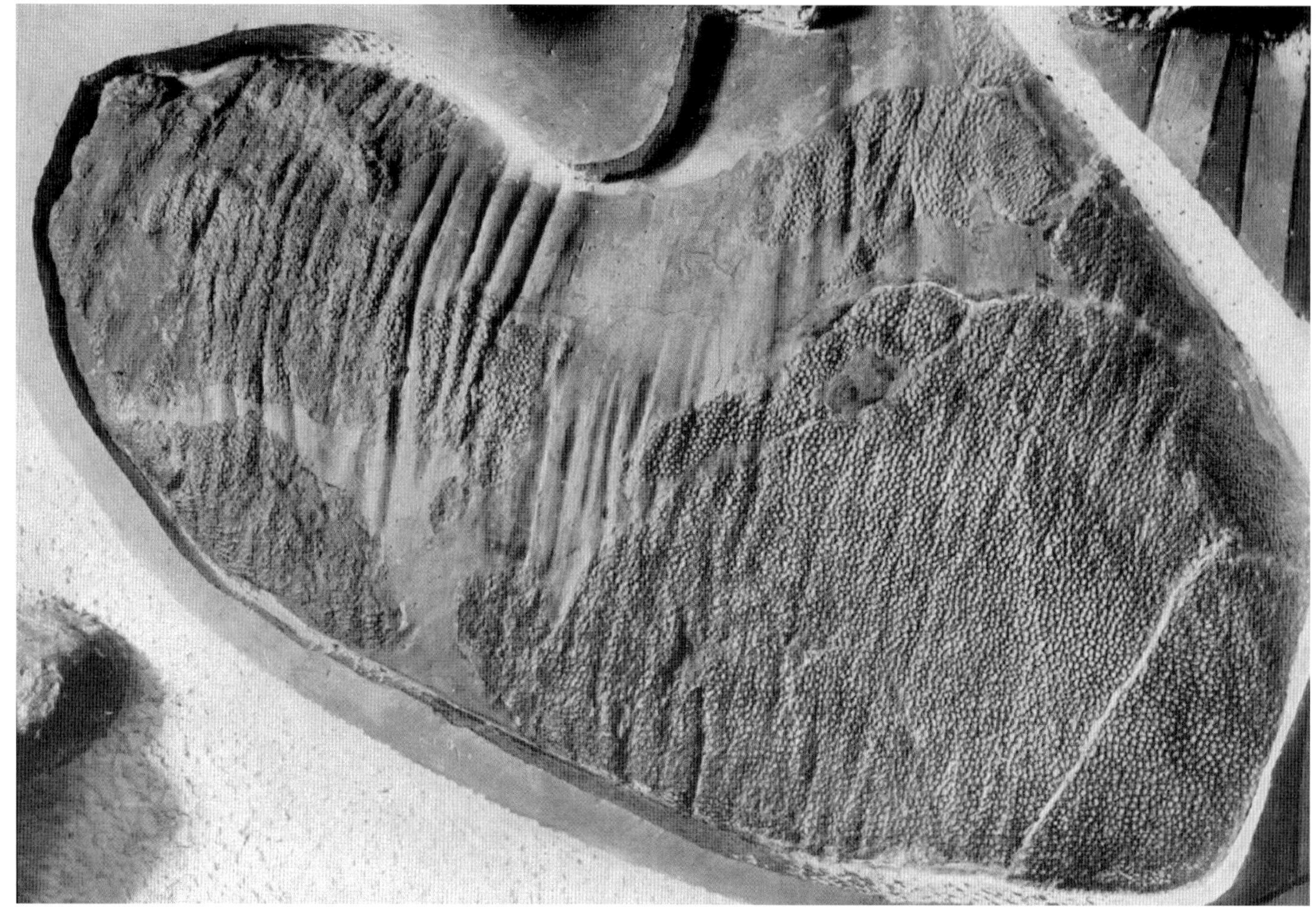

Skin impressions of *Corythosaurus* from a specimen collected in 1912.

Scientists have used powerful electron microscopes to examine the fibers on the tail of the theropod *Sinosauropteryx*, finding preserved microstructures called "melanosomes," which contain the pigment melanin. Different shapes to the melanosomes produce different colors in today's animals. By comparing modern melanosomes to the ones found preserved in the fossils, scientists were able to speculate on the colors of these extinct animals. The melanosomes in the filaments/fibers of *Sinosauropteryx* indicate chestnut to reddish-brown bands along the tail and possibly the back.[29]

Further study of the melanosomes in feathers from the bird *Archaeopteryx* showed a black color, somewhat like a crow or raven.[30] An additional study of the feathered bird *Microraptor* also showed melanosomes indicative of a black or dark blue color, but these melanosomes were more tightly packed, implying some degree of iridescence.[31] Spectacular preservation due to rapid burial in the Flood has allowed scientists to marvel at these exceptional and rare glimpses into the past.

At present, paleontologists cannot be certain of their color interpretations of dinosaurs, but we seem to be getting closer. Recent research with swimming reptiles from strata claimed by evolutionists to be between 55 and 190 million years old has revealed actual preserved skin tissue and melanosomes that indicate a brown-black color was dominant.[32] The ichthyosaur, mosasaur, and sea turtle included in this study all exhibited these same dark colors. Recall, the mere existence of these preserved soft tissues demonstrate that these fossils cannot be millions of years old, but are only thousands of years old, just as the Bible tells us (see chapter 3). Finally, several scientists in Canada are examining some newly discovered, well-preserved dinosaur skin fossils for melanosomes, so we may get some color indications in the very near future.

Animals like *Parasaurolophus* may have used their crest to make wailing sounds as described in the Book of Micah.

DINOSAUR SOUNDS?

What can science tell us about the sounds dinosaurs made? Recall from chapter 7 that the crests on Lambeosaurine dinosaurs were connected to the nasal cavities. When scientists blew air through anatomical replicas they heard a sound like a Scandinavian musical instrument called a *krumlur* or *krumhorne*.[33]

The Bible also provides clues to the sounds of post-Flood dinosaurs, and possibly gives us a true eyewitness account. Dr. Jim Johnson, chief academic officer at ICR, recently found evidence in the Book of Micah that some dinosaurs made a wailing sound. The prophet Micah said he would "wail and howl" and "make a wailing" like the dragons (Micah 1:8)." Dr. Johnson pointed out that dinosaurs were apparently known for this moaning-like wailing sound:

In fact, the nouns *tannin* and *tannim* (usually translated "dragons") derive from the Hebrew root verb tanah, which describes a sad, wailing lamentation (see Judges 11:40), meaning "to wail, moan, mourn, or lament." So the Hebrew noun *tannim* literally describes monsters known for wailing and moaning.[34]

Dr. Johnson concluded that this sad-sounding wailing fits the empirical finding from science. So we have both an eyewitness account and paleontological evidence that suggests hadrosaur dinosaurs emitted low-frequncy nasal sounds, like the tones from a krumhorn.

> ...I will make a wailing like the dragons, and mourning as the owls.
>
> —Micah 1:8; KJV

***Oviraptor* replica and eggs in the Senckenberg Museum in**

This chapter discusses some of the more recent discoveries and speculations on dinosaur behavior. Much of what we think about dinosaurs and how they interacted socially is unavailable in the fossil record. However, there are some things that can be gleaned about behavior from the study of footprints, egg nests, and even computer models.

Once again, all conclusions are a consequence of worldview. If you think dinosaurs are millions of years old, you will most likely interpret this data much differently from those of us that hold a biblical worldview. The data set is the same, however, and scientists are always making new discoveries. But alas, without a living dinosaur to observe, much of what is taught and published concerning dinosaur behavior is merely speculation.

DINOSAUR EGGS AND NESTS

Hadrosaur egg nest, collected in Hill County, Montana, on display at the Museum of the Rockies in Bozemen, Montana.

The first dinosaur eggs were identified by the Central Asiatic Expedition to Mongolia in 1923, led by Roy Chapman Andrews. Since that time, dinosaur eggs and nests have been found at over 200 sites around the world. The biggest dinosaur eggs are about 20 inches long and 5 inches in diameter, but most are much smaller, often less than six inches long. Dinosaur eggshells had been found in the USA since 1913, but were mostly misidentified or largely unrecognized until John R. Horner made his discovery of eggs and baby dinosaurs in Montana. He found the eggs at a site he named Egg Mountain for the *Maiasaura* eggs (a Hadrosaur) discovered there in 1978.

It should come as no surprise that, as reptiles, dinosaurs laid eggs. Yet there are some extinct marine reptiles that gave live birth, such as the ichthyosaurs. It is still not clear if all dinosaurs laid eggs or not. Although eggs contain all the food and sustenance for development of an embryo, they do need water, oxygen, and carbon dioxide to pass through the pores in the shell and therefore, need to be laid on dry land.

After careful study of the microscopic calcite minerals making up the eggshell, scientists have identified three basic eggshell designs common in dinosaur eggs — spherulitic, prismatic, and ornithoid. One of these, the spherulitic type, was laid only by non-theropod dinosaurs. Theropods, in contrast, laid eggs of the prismatic type and, more commonly, the ornithoid type.[1]

Many dinosaur egg nests are found with other nests of the same species. These colonial sites indicate that a type of social behavior, even if in a panic, was at play with the dinosaurs. John R. Horner has even determined the distance between adjacent nests is roughly equivalent to the adult body size of the species laying the eggs. In this manner, the mother dinosaurs would not crush one another's eggs in the process of laying their own, even if they rotated in a circle while laying the eggs.

Some nests are even found in the same area and at slightly different levels, separated by a thin layer of sediment. Evolutionists claim these findings as evidence of repeated colonial nesting, year after year, at the same site with local floods burying the nests each year. Creation scientists view these discoveries differently, as discussed below.

In 1997–1999, in the Auca Mahuevo area of Patagonia, Argentina, thousands of dinosaur eggs, and hundreds of nests from the same species of sauropod were found buried together in soft mud. The embryos inside were so well preserved that the designs in the skin were clearly visible. The investigating scientists found similar nests at the site in four separate horizons, with nearly 100 feet of sediment between the lowest and highest egg levels. They argued that these were deposited over at least four nesting "seasons."

However, creation scientists interpret this data much differently. In a Flood interpretation, these nests were likely laid within days or even hours of one another. Remember, sedimentary rocks don't tell time. It's all a matter of the interpreted speed of deposition. At this site, the depositional rate was so fast that each freshly laid nest level was engulfed before the eggs had a chance to hatch.

The pregnant sauropods had to keep laying their eggs, on each newly deposited surface, while trying desperately to avoid being buried themselves in the rapidly rising water and mud. They congregated on the "driest" pieces of land and made their nests near one another. Waves of water and sediment washed over each succeeding layer of nests, preserving the eggs and the embryos inside. Sediment separating the nest levels merely demonstrates the amount of deposition across this location that took place in a matter of days or even hours. Eventually, all the nests were buried catastrophically as the dry land disappeared under the Flood, along with some of the proximal adults, creating the fossils found today.

Most dinosaur eggs are found in some type of organized nest. Most dinosaur eggs are found in nests of 20 to 40, usually in some sort of circular shape, regardless of species. *Troodon*, a small theropod dinosaur, laid eggs in a circular clutch of 24 eggs in a shallow, bowl-shaped depression. There is a lot of speculation on whether a vegetative cover was placed over the eggs, but little hard evidence. Dinosaurs may have tried to cover the eggs but probably didn't have time as the Flood waters were rapidly approaching.

The American Museum of Natural History of New York, the same museum that conducted the Central Asiatic Expeditions in the 1920s and 1930s, was allowed back into Mongolia in the 1990s. They uncovered a spectacular fossil of an adult *Oviraptor* crouched over a nest of eggs. This find has been used as further evidence of parental care by dinosaurs and even incubation. However, there is no compelling evidence that the adult was incubating or brooding over the eggs. Fossil position doesn't tell us behavior. Fossils are found where they are buried, not necessarily where they lived or where they died, and if buried rapidly, it should not be a surprise to find some adults on or near the nests of freshly laid eggs, or even in the process of laying the eggs.

Scientists can "see" inside dinosaur eggs through CT scans (computerized tomography). I was able to have two hadrosaur eggs scanned in this way. One of the eggs showed nothing inside except fractures passing through the egg. The other egg showed something inside with a different density, implying possible embryotic remains were present. CT scans alone cannot determine whether or not a density difference represents a dinosaur. Scientists would have to dissolve part of the egg away to see for sure.

Many eggs, like the ones I scanned, were flattened from their original roundness due to burial pressure. Typically, the flattening causes fractures to pass through the egg, and any embryo inside dissolves away. Rapid burial alone doesn't guarantee the formation of a fossil.

Hadrosaur egg nest replica.

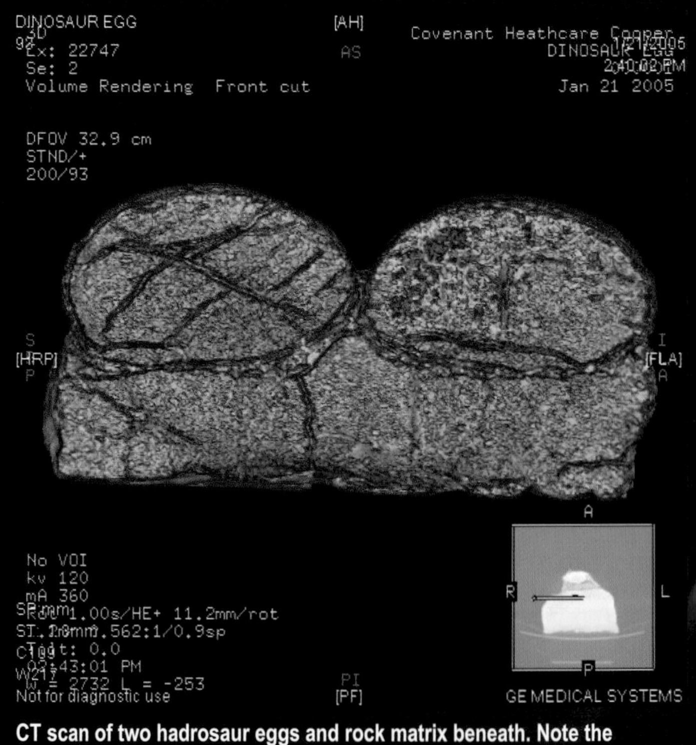

CT scan of two hadrosaur eggs and rock matrix beneath. Note the prominent fractures running through the left egg.

DINOSAUR HATCHLINGS AND PARENTAL CARE

Today, only about 10 to 15 percent of bird and reptile hatchlings make it through their first year of life. Dinosaurs likely had about the same survival rate. Whereas many birds mature to adulthood in less than a year, few dinosaurs grew to adult size in less than three or four years, and tyrannosaurs seem to have taken upwards of 15 years to mature. There is evidence some dinosaurs were altricial, meaning the hatchling was not fully developed when it emerged from the egg. And there is evidence that some dinosaurs were precocial, being fully functional from the day they hatched and requiring no parental care. Ostriches are precocial birds, and robins are altricial birds. Reptiles today are nearly all precocial, but some hatchlings stay near their parent briefly after hatching, like crocodiles. Others, like turtles, are literally, on their own.

John R. Horner has been on the forefront of hatchling and dinosaur parental care research. He has argued that the underdeveloped knee joints of *Maiasaura* (a hadrosaur) hatchlings indicates they were altricial and had to remain in the nest for some unspecified time after hatching. In contrast, Horner pointed to the hypsilophodontid *Orodromeus* (a smaller bipedal ornithopod similar in body structure to the *Maiasaura*) as evidence of precocial behavior. He found the leg bones in *Orodromeus* already fully developed in the hatchlings.[2] Both of these dinosaurs were found in nesting sites very close to one another and in the same rock layers in northwest Montana. Other paleontologists have argued, however, that the evidence for altricial dinosaurs is weak, noting that the knee of the *Maiasaura* was not any more underdeveloped than that of precocial alligators.

If the leg bone evidence is indeed sound, it appears some dinosaurs were precocial and others were altricial. God created some dinosaurs ready to be independent as soon as they hatched. Others He made briefly dependent on their parent(s). Diversity in parental care seems to be built in from the beginning.

Obviously, dinosaurs were most vulnerable when they were young. It seems likely that only a few percent ever made it to adulthood. Whether or not adult dinosaurs, like the *Maiasaura*, ever brought food back to the nests cannot be known. There is little debate, however, on the mode of preservation of these fossil hatchlings. They were rapidly entombed while in the nest by sediments carried by the great Flood.

An adult *Psittacosaurus* (parrot lizard) was found in China with 34 fully articulated hatchlings.[3] All of them were in upright positions, suggesting rapid burial while still alive. Scientists have used this find as evidence to support some degree of parental care in *Psittacosaurus*. The strength of this argument has been deflated, however, as claims have been made that the adult skull was merely added to the baby specimens, making

***Psittacosaurus* hatchlings**

Coelophysis

the parental care suspect in this case. The scientists agreed that the burial of these 34 hatchlings was rapid and catastrophic and likely caused by a sudden clastic (sand and clay) depositional event, encapsulating the juveniles instantly. Creationists point to this find as additional evidence in support of the Flood. The hatchlings were merely overwhelmed by the Flood waters and were preserved in life positions by the deposited sediment.

At least two juvenile *T. rex* partial skeletons were unearthed along with *T. rex* Sue. Although Sue had a skull more than 60 inches long, these two juveniles had skulls measuring 10 and 18 inches, respectively. In addition, there was a subadult *T. rex* lower leg found with the others. Mixed in the sediment around the tyrannosaurs were crocodile vertebrae and teeth, ornithopod bones, and a variety of small coelurosaur teeth. This likely does not represent the environment in which they lived. Instead, creation scientists interpret this site as further evidence of the catastrophic nature of the Flood, mixing many types of dinosaurs together, often tearing some apart and scattering their disarticulated remains in with other species. It may be that the tyrannosaurs were in close proximity, but considering the chaotic nature of the deposit, the evidence is not compelling.

It is becoming more and more common in secular publications and television documentaries to show baby tyrannosaurs with a coating of feathers or some type of downy covering on their skin. However, there is no evidence to support this notion. Feathered tyrannosaurs have never been found. The only "covering" scientists have unearthed for tyrannosaurs is a likely skin imprint, showing scaly skin. Dinosaurs are not birds and there are no feathered dinosaurs. There are only feathered birds.

At Ghost Ranch, New Mexico, hundreds of articulated adult *Coelophysis* (theropod) skeletons have been found buried with juveniles of all sizes of the same species.[4] Again, this is used by secular scientists as an argument in support of altricial care, but the proof is still lacking. The tremendous number of articulated skeletons at this site, found mixed together in a mass agglomeration of bones, can only be explained by a catastrophe that overtook these dinosaurs, burying them all together. Edwin Colbert thought this was some sort of local flood event. However, creation scientists correctly point to the great Flood as the explanation for this dinosaur gravesite.

HOW HARD AND FAST COULD *T. REX* BITE AND RUN?

Tyrannosaurus rex is likely the most famous predatory dinosaur of all. The *Jurassic Park* movies have pumped new life into its image as a savage, unstoppable menace. But how much of this is merely Hollywood hype, and how much is based on scientific evidence?

An adult, 5-metric-ton *T. rex*, if endothermic or warm-blooded, would have had to eaten the equivalent of an adult hadrosaur (duck-billed dinosaur) each week to supply its hunger needs. With teeth nearly a foot long (including the massive root), an adult *T. rex* could appear quite menacing. If ectothermic or cold-blooded, it would have required a lot less, maybe as little as a fifth or a tenth as much food, depending on activity levels. Paleontologists have matched up numerous tooth marks on dinosaurs to the dental patterns of the *T. rex* and its unique D-shaped teeth. Gouges from a large carnivore were even found on another *T. rex* toe bone, implicating T .rex as a cannibal.[5] Most scientists think this not surprising, as many animals did this as they scavenged for food. There are a few paleontologists, like John R. Horner, who think *T. rex* was exclusively a scavenger, as it likely couldn't run very fast as an adult and it had a brain shape that indicated a large olfactory lobe (nearly half its brain). He reasoned that *T. rex* could have used its nose to smell dead carcasses from great distances away. However, healed bite marks, attributed to an animal the size of a *T. rex*, have been found on both a *Triceratops* and an *Edmontosaurus* skeleton, indicating they survived a predatory attack.[6] Most recently, scientists reported finding the tip of a *T. rex* tooth embedded in the backbone of a duck-billed dinosaur.[7] The backbone had healed around the tooth fragment, demonstrating survival for some time after the failed attack. Other paleontologists have concluded that there wouldn't be enough dead prey available to feed large predators the size of *T. rex* if just a scavenger.[8] These studies suggest that *T. rex* was a predator at least some of the time. Most paleontologists agree that *T. rex* probably ate whatever it came across, as an opportunist, and not just a scavenger.

Studies of bite mechanics provide more data to support the notion that the *T. rex* was truly the "king of dinosaurs." Scientists in England used dynamic musculoskeletal models to simulate the bite strength of a *T. rex*. They found that it had nearly double the biting force of of an equivalent-sized alligator, finding values between 7,800 and 12,800 pounds of force.[9] Crushing any prey with this bite would have been easy!

Regardless of calculated bite strength and teeth marks, in God's original creation even *T. rex* was a vegetarian, as were all animals. In Genesis 1:30, God gives "every green herb for food" to the land animals and dinosaurs. It wasn't until after the sin of man and the resulting Curse that *T. rex* became a true meat-eater. Genesis 3:14 extended the Curse to every beast of the field (including dinosaurs).

Could an adult *T. rex* run fast enough to chase down prey and sink those teeth into it? Some scientists believe a younger *T. rex* may actually have been able to run, but as it got larger, was forced to slow down due to the increased mass. R. McNeill Alexander, from the University of Leeds, England, used computer models to conclude that a large adult *T. rex* had the mobility of "an asthmatic elephant" and could move at a maximum speed of only 5 mph. Therefore, he and Jack Horner believe *T. rex* adults were more likely scavengers, even stealing the kills of other dinosaurs.

A more recent study from the University of Manchester, England, using scale biomechanical models, determined that an adult *T. rex* could run at a top speed of a fast human, about 18 mph.[10] So it seems likely *T. rex* could chase down prey, and probably humans. Scientists will probably never be able to determine how fast *T. rex* could run, nor how hard its bite really was, without a live specimen to examine. But, based on what we do know, it's no wonder humans likely lived far from these types of dinosaurs before and after the Flood.

ATTACK AND DEFENSE MEASURES OF DINOSAURS

Predatory dinosaurs, the theropods mostly, are portrayed in movies as hunting in coordinated attacks. How accurate is this notion? Most of this, unfortunately, is speculation and is not testable science. The common consensus seems to be that smaller predators hunted in packs, similar to wolves or dog breeds today. On the other hand, most paleontologists think the larger *T. rex*–sized predators hunted alone, or in small groups, somewhat like lions today. These interpretations are based mostly on modern mammalian analogy, but also on some fossil evidence. Unfortunately, there is a limit to deriving behavior in life from dead fossils alone. Groups of small theropods, like *Coelophysis*, have been found fossilized together in groups of up to 40 or more individuals, showing they lived in close proximity.[11] Some larger theropods, like *Allosaurus*, have also been found in groups of over 40 in central Utah. Finally, the famous *T. rex* "Sue" was excavated out of rock containing the fragments of three other *T. rex* individuals, indicating tyrannosaurs lived in close proximity to one another also.[12] Paleontologists have also found footprint evidence showing theropod dinosaurs possibly traveling together and making parallel tracks, but not usually in large herds.[13] What are we to make of all this? No one is really able to tell how theropods hunted based on fossil evidence.

These groupings of predators may be partly a consequence of the Flood itself. The dinosaurs may have been "herded" together in their attempt to survive the rising Flood waters, gathering on the limited land still available prior to their demise. Fossils do not tells us if they truly lived in family units or hunted in packs, just that the Flood sediments buried many of them close together.

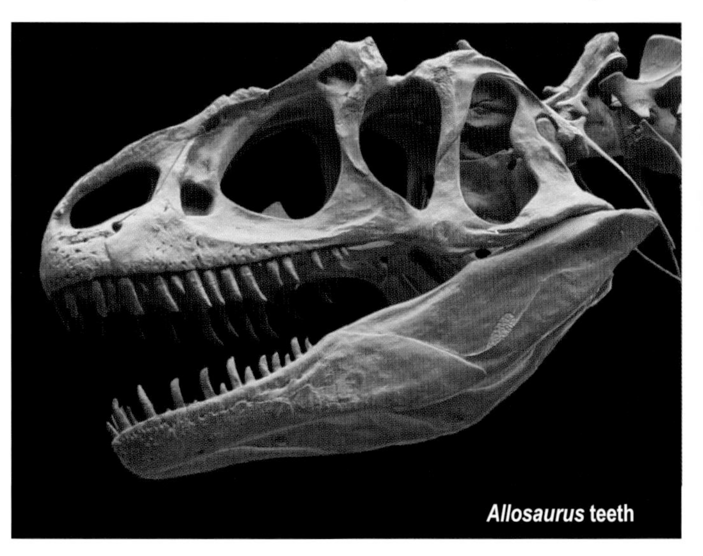

***Allosaurus* teeth**

Most predators were designed with sharp, serrated teeth, and possibly used them after the Curse for attacking in a shark-like fashion. These dinosaurs may have killed by causing a lot of bleeding to weaken prey. Other predators, like *T. rex*, were designed with much thicker teeth and may have killed after the Curse by tearing large bites out of prey.

Stegosaurus may have used its well-designed fan shape and plates to appear much bigger to ward off interested predators as it turned sideways, in a broadside fashion. Other plant-eating dinosaurs may have been able to run fast or dart quickly to avoid attacks. Some scientists have suggested that the sauropods (long-necks) used their necks or tails to strike at any would-be attackers.

Dinosaurs like *Ankylosaurus* were created with a protective covering of bony plates and a large tail club that may have been wielded at attacking predators. This large mass of bone has been shown capable of breaking even a *T. rex* leg bone. Stegosaurs were provided with tail spikes that could have also helped deter an attacker. Even the sharp, designed horns on ceratopsians may have been used to lunge at attackers when necessary.

Footprint evidence and fossil death assemblages often show extremely large herds of plant-eating dinosaurs, sometimes more than 10,000 buried together. Living in large herds was probably an effective deterrent to attack from predators. Young dinosaurs were likely most susceptible to attack, as they were slower and didn't possess the protective clubs, spikes, horns, and tough skin of a mature adult. Size must have been a deterrent also. Dinosaurs probably became progressively more immune to attack as they grew larger and larger.

Stegosaurus

SOCIAL MARKINGS AND GROUP BEHAVIOR OF DINOSAURS

Although behavior of extinct organisms is difficult to extract from fossils, there are some skeletal features of dinosaurs that may have been used for socialization and/or status in a herd setting. Many of the head crests on adult lambeosaurine dinosaurs, such as *Parasaurolophus*, were designed to connect to the nasal passages and were likely used to make sounds. These head crests came in a variety of sizes and shapes. Based on studies of these tubes, they seem to have been designed to make resonating sounds, useful for communication to members of the herd, such as making warning calls and/or mating calls. The different sizes of the crests may also be indicative of males and females of the same species or they could be merely differences within the "kind." Unfortunately, we don't know which are males and which are females, and the differences are too often used to justify a separate species instead of sexual dimorphism.

Social rank or family recognition may have played a role in the wide variety of horn sizes, shapes, and frill ornamentation in the ceratopsians. Some of these variations likely helped establish social status in the herd or, alternatively, could have been due to differences in the sexes. Some paleontologists have noted that hatchling ceratopsians all looked much the same. It wasn't until they began to mature that the horns and frills took on distinctive character, leading to the interpretation that these differences were for recognition within family units.

Creation scientists believe that horn and frill differences in many of the ceratopsian dinosaurs do not necessarily establish separate species. They are explained as variations within the same species or kind, just as dogs have different breeds today. Evolutionary scientists are prone to name any noticeable difference a new species, artificially inflating the number of dinosaurs.

Some of the dinosaurs we find buried in mass death assemblages may be an artificial consequence of last minute "herding" as the dinosaurs scurried to find the remaining land during the rising Flood waters. Thus, all arguments for socialization using fossil assemblages may be flawed. However, the largest dinosaur death assemblages, with thousands of the same species, suggest more strongly that herding behavior was occurring before the Flood.

Dinosaur footprints are also cited as evidence of socially organized herds, as many similar trackways are oriented in the same direction. However, footprint data may also be strongly influenced by the "herding" effect of the rising Flood waters, causing many to travel the same limited paths in an attempt to escape. We will discuss more about this in the next section.

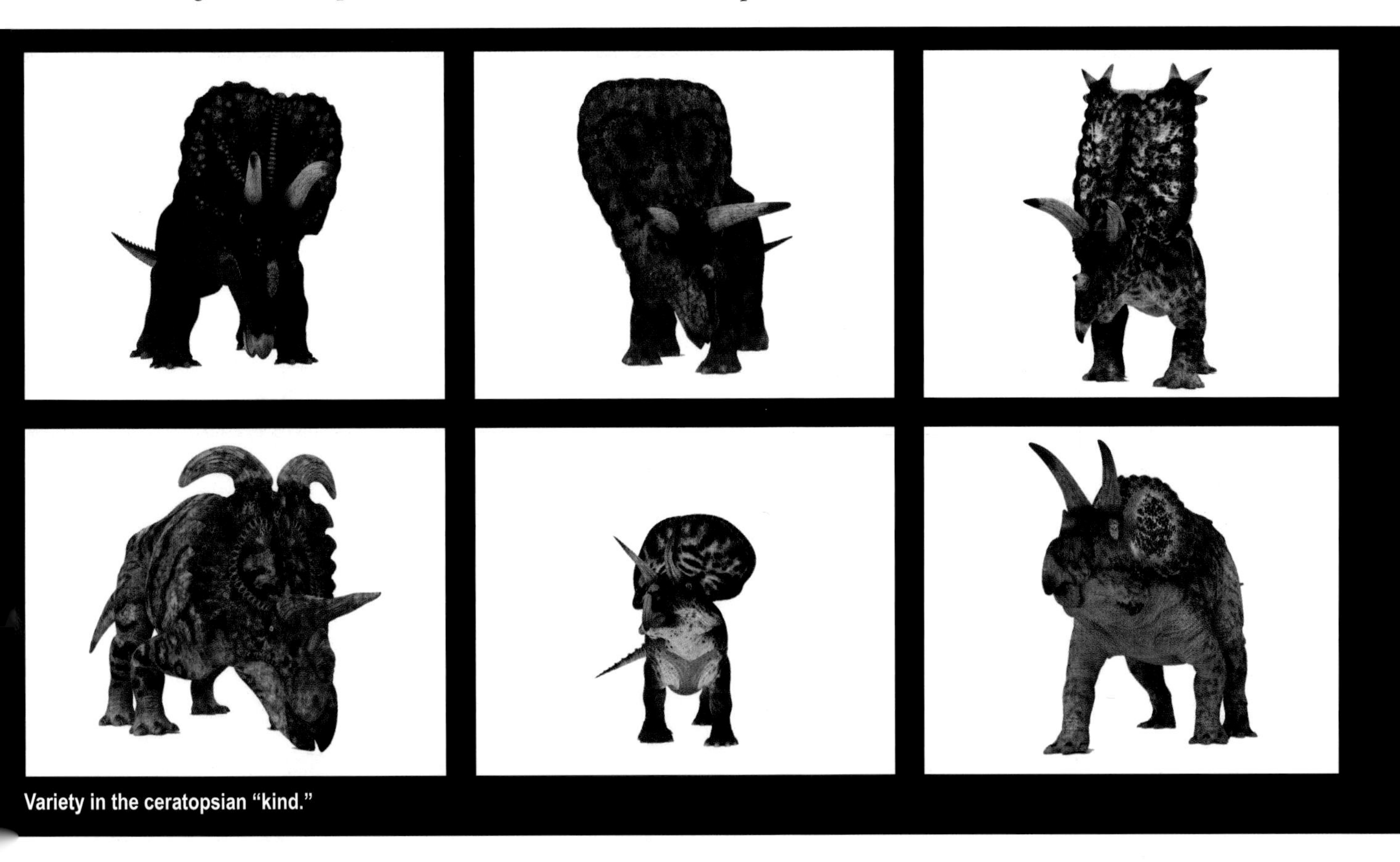

Variety in the ceratopsian "kind."

DINOSAUR FOOTPRINTS

Many thousands of dinosaur footprints have been found at about a thousand locations worldwide, and more are found each year. Every continent has yielded dinosaur footprints except Antarctica, where they will probably be found in the future. Creation scientists point out that footprints shouldn't be so commonplace under normal circumstances, as the preservation of a footprint requires special conditions, just like preservation of bone fossils. Footprints don't last more than a few days in most environments today. Special conditions, however, were prevalent during the Flood of Noah, preserving many prints and tracks as evidence of catastrophic burial.

The first discovery of a "dinosaur" footprint was by Pliny Moody in 1802 while plowing a field in New England. Because dinosaurs were not known at the time, these were considered to be from a large bird and the significance of his discovery was largely unnoticed until much later. Some of the earliest studies of dinosaur footprints were done by Edward Hitchcock in the mid-19th century. Unfortunately, most scientists of the time were caught up in the search for dinosaur bones, and footprints were neglected until recently.

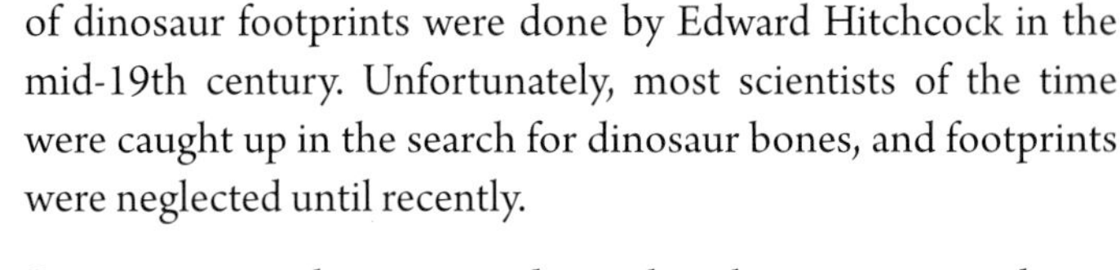

Some scientists have pointed out that there appears to be an overrepresentation of predator (theropod) footprints when compared to the predator bone record. In rocks, bones of predators make up only 3 to 5 percent of the dinosaur fossils. Tracks at the Purgatoire site in Colorado, in Jurassic system rocks, show a 60 percent theropod (predator) to 40 percent sauropod (prey) distribution. And tracks within the Dakota Group (Cretaceous system) near Denver, Colorado, at Dinosaur Ridge exhibit proportions of 25 to 30 percent predator to 70 to 75 percent prey.

Why the discrepancy between dinosaur track data and the bone data? Some have suggested the higher activity levels of the theropods as a possible solution. This solution makes even more sense if considered in a Flood context, however. During the Flood, the theropods may have used their higher speed and agility levels to disproportionately escape the advancing Flood waters and leave more tracks, compared to the slower and less agile plant-eaters. This may be one reason why the proportion of theropod bones does not match the proportion of theropod tracks.

Many dinosaur footprints, like those found in British Columbia, show individuals traveling in the same direction. This suggests that some may have traveled in social groups, and/or also implies they were moving in a coordinated manner away from something. Creation scientists think these unidirectional track ways were formed during the early to middle stages of the Flood as dinosaurs traveled away from the danger of the rising water and tsunami-like waves.

Further support for this interpretation is demonstrated by dinosaur footprints in rippled sandstone layers and between rippled layers. Tracks with rippled layers demonstrate dinosaurs were wading through shallow, actively-moving water. The coincidence of trackways and ripples is strong evidence that these dinosaurs were trying to escape from rising Flood waters.

Sandstone rockface containing an unidentified dinosaur footprint in bottom left, near Severn estuary, United Kingdom.

Sauropod dinosaur tracks at the Purgatoire River site, Colorado.

DINOSAURS HAD TO SWIM OUT OF NECESSITY

There are even more indications that many dinosaur tracks were made in shallow water. Discoveries of swimming dinosaurs have been reported from several locations in Europe, Australia, North America, and China.

Researchers studying dinosaur footprints in China's Sichuan Province reported their discovery in the *Chinese Science Bulletin*. Their article documents several different types of dinosaur track ways, including two theropods (meat eaters), one sauropod (long-neck), and four ornithopods (duck-bills) all traveling in the same direction at about the same time.

The scientists determined at least one of the theropods was partly afloat, as much of its 50-foot (15 m) trackway shows just tip-toe scratches and claw scrapings. A second theropod, walking in the same direction and on the same horizon, was apparently wading on the bottom, as its footprints were more complete. The secular scientists interpreted that the tracks were made as the dinosaurs crossed a river, possibly during a flood event.

Both theropods in this case were estimated to have hip heights of just less than 3 feet (1 m). The scientists concluded that the pair of theropods were simultaneously swimming and wading as rapid fluctuations in water depth were occurring.

In addition, the parallel trackways of the sauropod and the ornithopods at the same site also indicated they were wading through water. The hip heights of these animals were slightly higher than the two theropods, giving them no need to swim. They were merely walking through the water.[14]

Creation scientists have suggested many animals swam to escape the Flood waters.[15] These trackways confirm that water overwhelmed the dinosaurs quickly and that they swam for their lives for as long as they could, as described in the Book of Genesis. The evidence of rapid fluctuations in water level is also expected as tsunami-like waves quickly overwhelmed the dinosaurs during the Flood. However, the evolutionists can only resort to trying to explain their observations with a local river flood. The consistent direction of the trackways also shows that all the dinosaurs were traveling in the same direction at the same time trying to escape the approaching Flood water.

A similar situation was reported at the Australian trackway site in Queensland, Australia. Here, again, scientists found evidence of the same species of dinosaurs walking and wading, indicating similar rapid fluctuations in water depth.[16]

Finally, a mud-trapped herd of over 20 ornithomimid (bird-mimic) juveniles was unearthed in Upper Cretaceous system rocks (later Flood) in Mongolia. Scientists discovered that these dinosaurs were walking through water-saturated, freshly deposited, muddy sediments, sinking in so deeply their feet and tails penetrated into underlying horizons as they struggled to survive.[17] The skeletons were all found in a single bedding plane and showed no signs of weathering, indicating a rapid burial. They also noted that most of the dinosaur skeletons were strongly aligned and laid out parallel to one another.

The obvious Flood explanation for this assemblage was left out of the secular paper. It is yet another example of the mounting evidence in support of the worldwide, catastrophic Flood. The footprint evidence shows that these dinosaurs traversed across freshly deposited, earlier Flood sediments, saturated with water, looking for an escape route. As the water and mud levels continued to mount in the Flood, we find dinosaurs such as these sinking in mud, struggling to survive, before succumbing to the inevitable and rapid burial.

The evidence for the Flood is found all over the world. Trackways of swimming dinosaurs and dinosaurs mired in muddy sediments are found on many continents. This data further verifies the global extent of the cataclysmic Flood. Dinosaurs ran away and swam for as long as they could, but eventually, all were buried in the overwhelming power of God's judgment as reported in Genesis.

CALCULATING THE SPEED AND SIZES OF DINOSAURS FROM FOOTPRINTS

Tracks are the primary sources for estimating dinosaur walking and running speeds. Unfortunately, we will never know if our speed estimates are truly correct, leaving room for many future disagreements.

Dinosaur footprint data shows many of them herded together in limited spaces, seeking refuge from the rising and assuaging Flood waters. Sometimes we see multiple species in the same area, milling around, looking for the highest ground, following their instincts to survive. As scientists, we can even use these tracks to estimate how fast they walked, and in a few cases, how fast they ran. But before we begin analyzing dinosaur footprints for speed, we need to go over some terminology. "Stride" is the distance between the same spot on consecutive footprints made by the same foot. Usually heel-to-heel is best. "Step" is the distance between consecutive footfalls (right-to-left or vice versa) of either the front feet or the back feet of quadrupeds, or between the back feet of bipeds.

A simple way to estimate dinosaur hip height was determined by R. McNeill Alexander.[18] Alexander found hip height can be approximated by simply multiplying the foot length by four. This works pretty well for most mammals and is extrapolated to bipedal dinosaurs, and for the back feet of quadrupedal dinosaurs. Hip height allows paleontologists to better estimate the size of a dinosaur using only footprint data.

Alexander further proposed a method of separating dinosaur "walking" tracks from dinosaur "running" tracks. He suggested that walking is indicated if step is less than four times the foot length, and running is indicated if step is greater than four times the foot length.

Finally, Alexander developed an equation using regression analysis to estimate dinosaur speed:

$$\text{Speed (in m/s)} = 0.25 \text{ x (gravitational acceleration)}^{0.5} \text{ x (stride)}^{1.67} \text{ x (hip height)}^{-1.17}$$

Note: The equation requires that stride and hip height be measured in meters!

All anyone needs to estimate dinosaur speed is the stride and hip height, which are easily determined from footprint data. For example, if the stride of a quadruped is 3.1 meters and the hind footprint length is 0.6 meters, the hip height can be estimated at 2.4 meters (4 x 0.6), and the speed is:

$$\text{Speed} = 0.25 \text{ x } (10\text{m/sec}^2)^{0.5} \text{ x } (3.1\text{m})^{1.67} \text{ x } (2.4\text{m})^{-1.17}$$
$$\text{Speed} = 1.88\text{m/sec}$$

Adult and juvenile *Iguanodon*-like tracks in Colorado. Note the missing prints marked by circular holes that were taken before the site was protected.

To convert m/sec to mph, multiply your answer by 2.2, as 1m/sec = 2.2mph. Our example indicates the dinosaur was walking at about 4.1 mph, a brisk walking speed.

Is there a way to determine which dinosaur species left the footprints? Unfortunately, we can only estimate the general type of dinosaur, whether or not it was bipedal or quadrupedal and often an approximate kind, such as a sauropod or ornithopod.

Footprint data can also be used for other information. Dr. Martin Lockley developed a graphical technique useful for estimating the body size of quadrupeds from footprints.[19] The overall shoulder-hip length (body size) of a quadruped can be determined by measuring the distance between the midpoint of the front footprints and the midpoint of the hind footprints. Using the four footprints of the front and back feet, measure the midpoints between each pair (the center of the shoulders), and then the distance between those two midpoints. This is called the shoulder-hip length. To do this, care must be taken to use the four footprints that the dinosaur would have had on the ground at the same time. The resulting shoulder-hip length is the trunk length. From this, an easy comparison can be made to known dinosaurs with the same trunk length, giving us an estimate of overall body size and even weight.

In the movies, we see many different portrayals of speed for the same dinosaurs. Estimates for the running speed of *T. rex* in the first two *Jurassic Park* movies varied considerably. Why aren't there better estimates for *T. rex*? For one thing, there aren't found many *T. rex* footprints. Second, there are few examples of running dinosaurs of any variety. Most footprints show dinosaurs were walking at "cruising" speed.

Three closely spaced sets of tyrannosaur tracks (likely *Albertosaurus* or *Daspletosaurus*) were discovered in British Columbia that showed a common direction of travel.[20] Significantly, these were the first multiple tracks of tyrannosaur prints ever reported, as most prior discoveries were just single footprints. Analysis of the stride and foot length suggests they may have been traveling between 4 and 6 mph. They didn't appear to be running, but were cruising along at a pretty fast pace. Again, these dinosaurs were likely trying to move to high ground and away from the advancing Flood waters.

***Iguanodon*-like prints and a small theropod track in foreground at Dinosaur Ridge, near Morrison, Colorado.**

Theropod tracks near Glen Rose, Texas.

DINOSAUR FOOTPRINTS ALONG THE PALUXY RIVER, TEXAS

Dinosaur footprints found along the Paluxy River are preserved within the Glen Rose Formation, dated by secular methods as between 105 and 115 million years ago (Cretaceous system). Many of the tracks today are preserved under the protection of Dinosaur Valley State Park. News of the first prints at the site were first published by Roland T. Bird in 1938, while he was an assistant to Barnum Brown at the American Museum of Natural History in New York. Bird believed the tracks were from a brontosaur-type dinosaur. Paleontologists now think these were more likely brachiosaur-type prints, finding skeletal remains of the latter in the same rocks elsewhere in Texas.

Some of the trackways at Paluxy show about 12 brachiosaur-type prints and three theropod-type prints all walking parallel to one another. The three theropod trackways seem to have been made a little bit after the brachiosaurs had left their prints, as they often step onto the sauropod tracks.

Dr. John Morris, from the Institute for Creation Research, has personally studied and written about the dinosaur footprints along the Paluxy River near Glen Rose, Texas. These fossils have become famous because many have claimed to have found human footprints with the dinosaur footprints. Dr. Morris examined the supposed human footprints and nearby dinosaur footprints over a period of several years. His conclusion was that the footprints appear to be three-toed, or theropod footprints, and not human prints after all. He noticed that what initially appeared to be a human footprint, over time, often erodes downward until three toes appear, similar to other theropod prints in the area.

The best human-like footprints come from the "Taylor Trail" at Glen Rose, where over 20 steps were found in succession. However, Dr. John Morris has found that even these tracks transformed with erosion into clear, three-toed dinosaur prints.

Dr. Morris has pointed out that at the time the prints at Paluxy were made, the Flood waters had transgressed across the continent several times, "depositing and eroding with unimaginable fury." He found it difficult to imagine that any humans could have survived the Flood catastrophe up to this point. Obviously, a few "hearty" dinosaurs had survived and were still wading through the freshly deposited lime mud, only to be buried in sediment at some point a bit later. Dr. Morris also found the presence of a "plethora of carbonized wood chips" mixed within the Glen Rose Formation. He believes toppled trees became broken into small pieces by the fury of the advancing waters of the Flood and were buried within the rapidly deposited limestone layer.

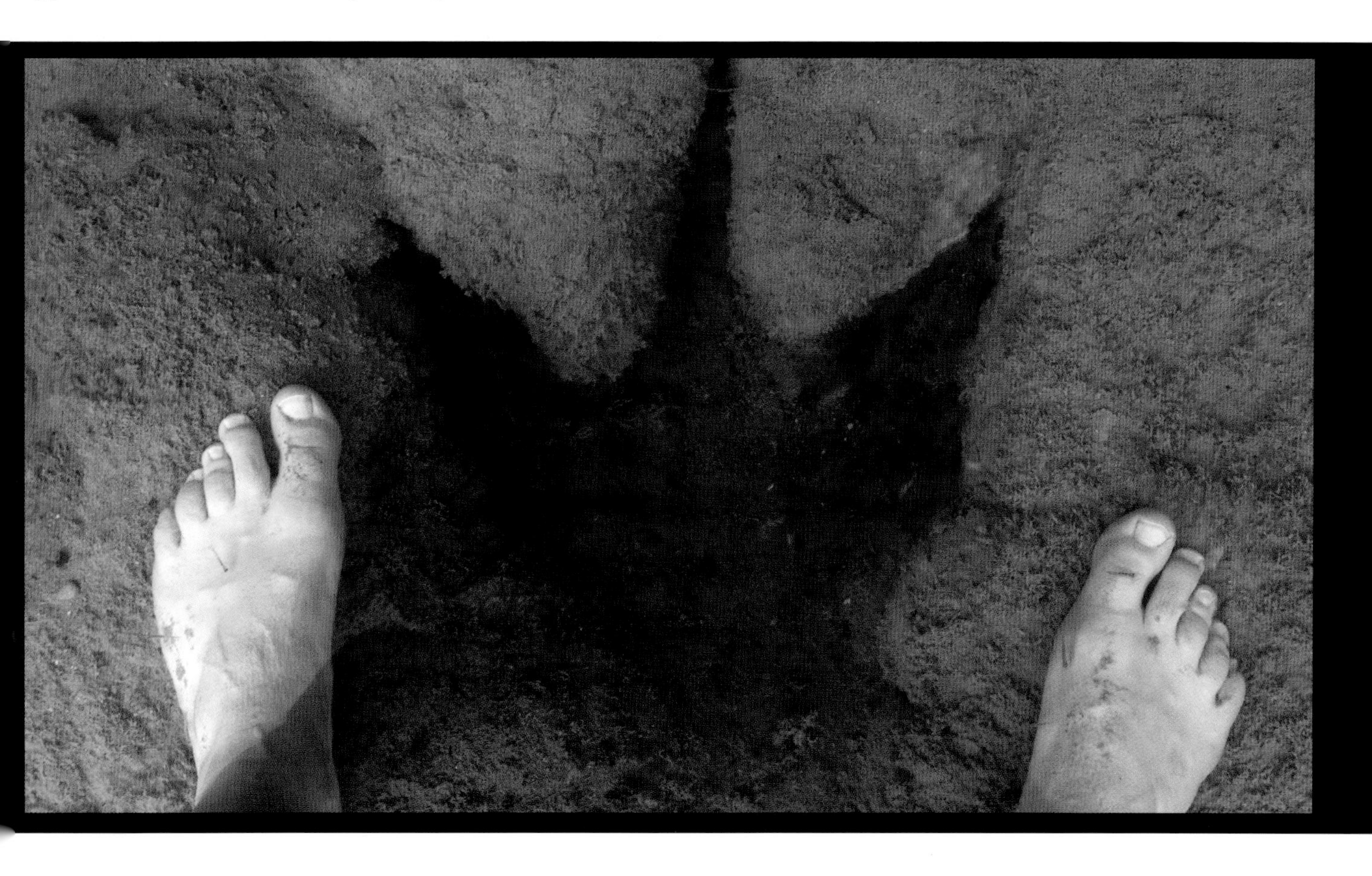

DINOSAUR FOOTPRINTS AT DINOSAUR RIDGE, COLORADO

Many dinosaur footprints are located just west of Denver, Colorado, along the Dakota hogback of the Front Range at a site called Dinosaur Ridge. The site is now protected and has been designated a National Natural Landmark. Although many of the prints are in a different rock formation, the site is near the original Morrison Formation bone quarries where *Stegosaurus* and *Apatosaurus* were first excavated in 1877.

Dinosaur Ridge is part of a larger exposure of dinosaur prints that extends from south of Denver at Castle Rock, Colorado, all the way north to Boulder, Colorado, a total distance of about 45 miles. Additional track sites have been found that extend even farther north and south, for hundreds of miles, from Montana to New Mexico. This large trail of track sites has been called the "Dinosaur Freeway."

Various types of prints have been found at Dinosaur Ridge, including tracks from several different rock layers. Most of the tracks are from the Dakota Group, a Cretaceous system deposit. Other tracks come from the adjacent, but slightly earlier and deeper, Jurassic Morrison Formation. In addition to dinosaur prints, amphibian prints, crocodile-like prints, and even bird prints have been found along this "freeway."

The Dinosaur Freeway exists because it's likely this was the highest ground and least flooded area, following deposition of the earliest Flood sediments across the continental USA. Early Flood sediments are quite thin in the Rocky Mountain vicinity in general. Dinosaurs trying to escape the rising waters would have naturally migrated to these highest, least-flooded locations. Many probably came there at different times, leaving tracks in different layers, possibly separated by just days at most, trying to avoid being overcome by water. This may be one of the reasons the dinosaurs were able to survive to this point in the Flood, as will be discussed in chapter 14. The "freeway" is a result of an exposed narrow strip of land that was temporarily available as Flood waters oscillated prior to the final advance.

The most easily accessed dinosaur prints are found in the Cretaceous Dakota Sandstone, dated by secularists at 100 million years old. These consist mostly of *Iguanodon*-type (duck-billed) dinosaurs. Some of the prints show the footpads and foot skin details of these dinosaurs. *Iguanodon*-like prints at Dinosaur Ridge show evidence of both young and mature dinosaurs traveling together. Small theropod prints are visible near many of the *Iguanodon*-like prints. Scientists think these prints are from an ostrich-like ornithomimid. These prints may have been made at nearly the same time as the *Iguanodon*-like prints or within a few hours or days of each other.

Many of the footprints in the Morrison Formation at Dinosaur Ridge are visible only in cross section, not from the top down like most prints. Brontosaur-type prints are up to three feet in diameter and "sink" into the soft layers below, indicating the layers were freshly deposited. Some of the bones found within the Jurassic system rocks of Dinosaur Ridge contain considerable amounts of radioactive uranium and can be sensed by a Geiger counter. The uranium was transported from nearby uranium sources into the bones as groundwater percolated into the pore openings in the center of the bones.

"Brontosaur Bulges" in the Morrison Formation at Dinosaur Ridge. Author's daughter for scale. These bulges demonstrate that the sediments were still wet and soft before the sauropod dinosaur contorted the layers. The flat layers above indicate the prints were subsequently buried rapidly, preserving the "bulge."

Tyrannosaurus rex and an ankylosaur

CHAPTER 12:
DINOSAUR ENDINGS AND EXTINCTION

EXTINCTIONS: WHAT ARE THEY REALLY?

Uniformitarian scientists have identified five so-called great extinctions in the Phanerozoic Eon, encompassing the rocks from the entire Flood record. The table below summarizes these five events by the purported percent extinction by species and by purported vertebrate extinction rate. Notice that the vertebrate extinction rate through time, by families per million years, shows different values compared to the species extinction rates. This demonstrates some of the statistical bias that can be used, depending on if you're talking extinctions by species or families or by rates of extinctions vs. extinction percent. The extent of the so-called Permian-Triassic extinction, or any extinction, depends on what is being counted. Secular scientists also assume their absolute dates are real.

The Great Five Extinctions

Extinction Event	Species Extinction (percent)	Vertebrate Extinction Rate (Families per Million Yrs)
Cretaceous-Tertiary	70	16
Late Triassic	76	10
Permian-Triassic	96	15
Late Devonian	82	10
Late Ordovician	85	20

In terms of factual data, none of these extinctions are real events. These merely represent changes in the fossil content from one layer to the next. The more sudden the change, the more likely it is claimed to be an extinction event. Secular scientists calculate the amount of change in the fossils from layer to layer, add in their assumed absolute dates for the layers, and then determine a so-called extinction rate with time. It is all a process steeped in evolutionary dogma.

The five great extinctions do, however, represent Flood event boundaries where rapid changes in sedimentation style have occurred. Comparing the five great "extinctions" to the boundaries of the six accepted megasequences, we find a rough correlation, at least with some of them. A few of the "extinctions" are roughly coincident with the

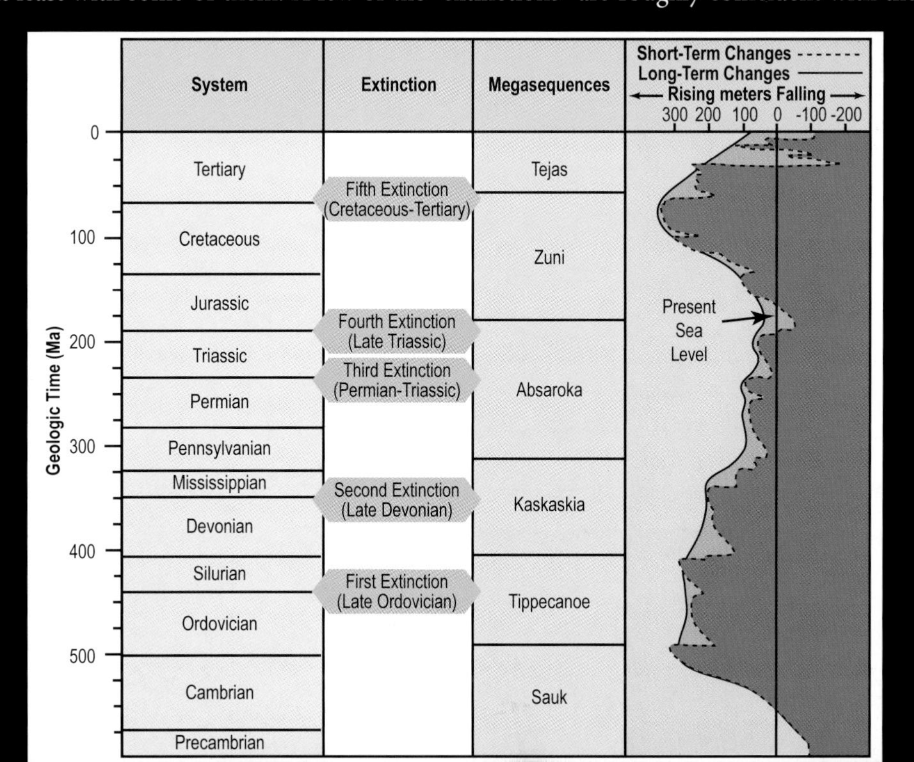

Secular chart showing the geologic time scale, the interpreted changes in relative sea level, and the six megasequences. The five so-called great extinctions are also shown. Diagram modified from the original provided by Dr. Andrew Snelling, *Grappling with the Chronology of the Genesis Flood* (Green Forest, AR: Master Books, 2014).

megasequences. Others fall in the middle of a megasequence. But even these often coincide with other accepted, minor extinctions. Admittedly, this is not a straightforward correlation, as both the Late Triassic and the Permian-Triassic extinctions fall within the Absaroka megasequence.

However, because the megasequences are defined on the basis of major shifts in rock type, it should be no surprise that some of these changes correspond to rapid shifts in the fossil content also. The fossils deposited are dependent on the tectonic forces at work, currents, waves, and the height of relative sea level at a particular point in the Flood. The fossil content is also dependent on the amount of erosion that took place between megasequences as the sea levels fluctuated. This may explain why some of the "extinctions" fall in between megasequence boundaries and others coincide with the tops and bottoms. So what looks like an extinction event to an evolutionist is really the result of a sudden shift in sedimentation pattern. The more dramatic the change, the more likely it is identified as a major "extinction."

Dinosaur fossils are found only in the fourth and fifth megasequences (from the bottom up), for the most part, known as the Absaroka and Zuni megasequences. It was while these two sequences were being deposited that water levels became higher and tectonic forces became greater, affecting the levels of sedimentation and the style of deposition. The dinosaurs were finally trapped by this more intense activity and entombed in sediment. There will be more discussion on this below.

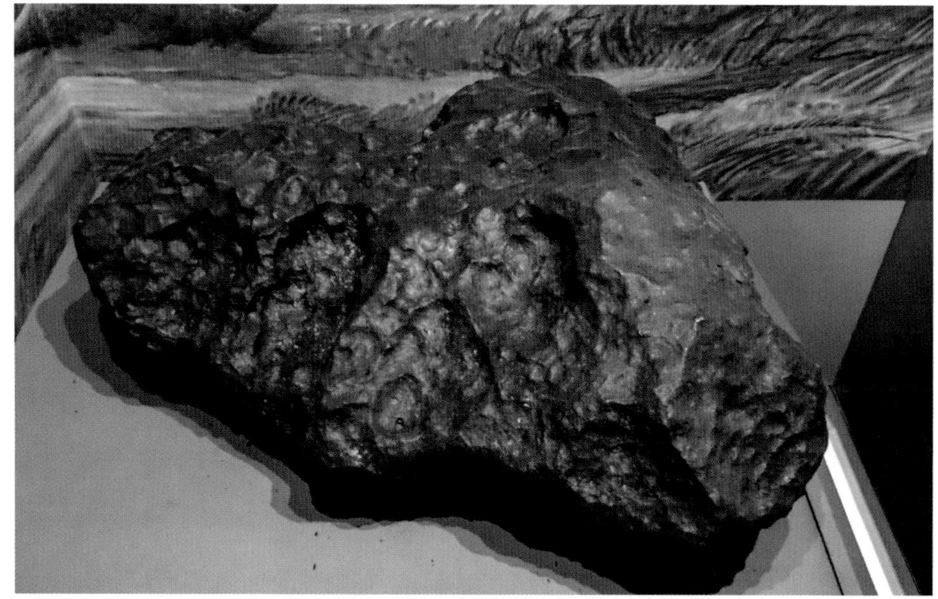

An exhibit of an asteroid rock believed by many secular scientists to have caused the dinosaur extinction.

DINOSAUR "EXTINCTION"

One of the greatest mysteries of all time concerns the extinction of dinosaurs. This extinction is commonly called the "K-Pg" event because it occurs in rocks near the Cretaceous-Tertiary (now called the Paleogene or Pg) system boundary. The "scientific" consensus on the absolute date of the K-Pg extinction event is approximately 66 million years before present. There have been well over 65 serious, published, ideas attempting to explain this extinction. Some of these ideas include medical problems, overspecialization, competition with mammals, plant changes, climate changes, sea level changes, increased volcanism, and the most popular, extraterrestrial impacts.

Contrary to what you might hear, many dinosaur paleontologists do not feel an asteroid impact is the primary answer to this mystery. They believe an impact, if it happened, only had a minimal extinction affect. Dr. Robert Bakker thinks that foreign diseases carried across land bridges, from continent to continent, killed the dinosaurs. Jack (John) R. Horner spoke about extinction at a lecture at Central Michigan University in April 2001. He stated that the asteroid theory of extinction is a "good example of taking an idea too far," noting that this theory is all based on negative evidence. Horner frequently reminds his audiences that no dinosaurs have been found within the last few feet of rock just prior to the K-Pg boundary, implying that dinosaurs were already extinct prior to the Tertiary Period. He believes that dinosaurs went extinct from simultaneous sea level changes and climatic shifts. He readily admitted, "I don't care, I'm just glad they're gone. I want to go out for walks at night."

THE SECULAR ASTEROID THEORY

The asteroid hypothesis is based on the "accidental" discovery of higher-than-expected levels of iridium (Ir) within the K-Pg boundary layer at Gubbio, Italy.[1] Similarly high levels of Ir were subsequently found at over 100 sites worldwide at the K-Pg boundary. Iridium is a rare earth element, found in meteorites, but not common on earth, and so it is thought to have come to earth from space. Many have concluded that this Ir-rich layer was deposited after an asteroid-like body struck the earth, scattering Ir-rich dust all over the globe.

Further study of the K-Pg boundary layer found tiny crystals of "shocked" quartz at many of these same locations.[2] Shocked quartz can only form under extremely high pressure conditions, like impact sites, further strengthening the evidence for an impact at the K-Pg. In 1992, even microscopic diamonds were reported from the K-Pg layer in Alberta, Canada.[3] It is thought that the diamonds are fragments of impacts, as they are "cleaner" and contain less nitrogen than traditional mantle-derived diamonds.

The presumed site of the K-Pg impact was located in 1991 near the northwestern edge of the Yucatan Peninsula, Mexico.[4] The site consists of a large, circular gravity anomaly known as the Chicxulub Crater. It is 110 miles (180 km) in diameter, allowing room for a 6-mile-wide (10 km) asteroid to have created the dimensions of the crater. There is, however, disagreement on the exact size of the crater. Some have reduced it to only 60 miles in diameter, which would make the impacting body much smaller.

The presumed crater is not visible at the surface because it is covered over by younger sediments. Deeper rocks at the site (Cretaceous system and older) are apparently the source of the circular gravity signature. Oil wells drilled into the crater have recovered cores of broken and brecciated rocks in the pre-Tertiary section, confirming the likelihood of an impact. Advocates for this hypothesis refer to the Chicxulub Crater as the "smoking gun," and use it as confirmation that an asteroid hit the earth. However, there is still not enough melt rock at Chicxulub to justify a large impact and virtually no iridium has been found in any of the wells at the site.

Secular scientists even calculated a 26- to 28-million-year, best-fit cycle to the extinction record across geologic time.[5] This apparent pattern is usually explained as the result of our solar system passing through an asteroid belt, or something similar, every 26 to 28 million years, causing an increased likelihood of impacts and the cyclic extinctions. It is, however, not based on much real scientific evidence at all, as we will see below.

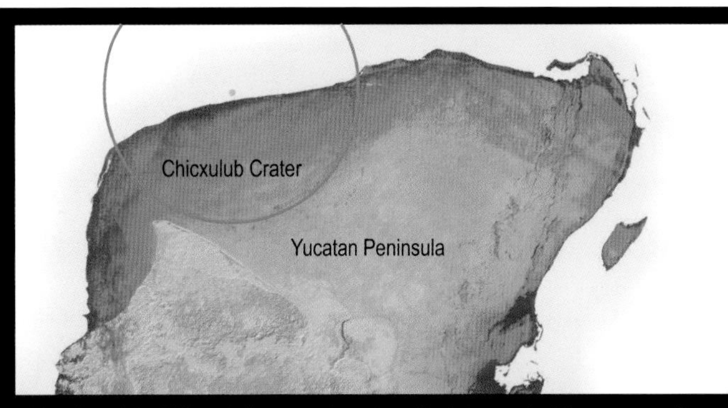

110-mile wide Chicxulub crater is primarily interpreted from a gravity survey.

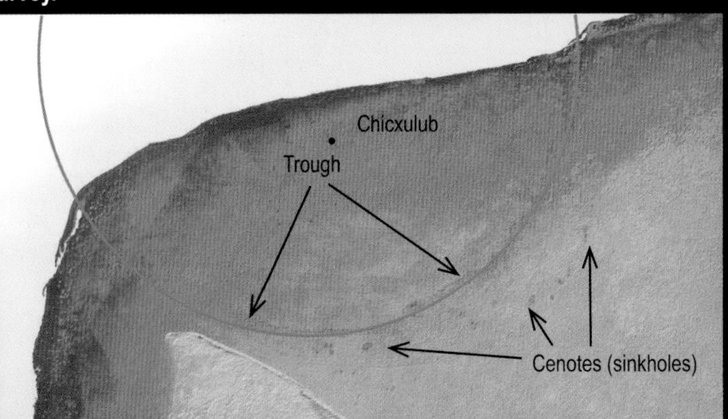

The crater is not visible at the surface.

Core sample from the rocks below the surface showing broken debris (breccia) purported to be from the Chicxulub impact.

SECULAR ARGUMENTS AGAINST THE ASTEROID THEORY

Other secular authors have argued that substantial iridium (Ir), and even shocked quartz can come from the mantle via volcanic eruptions.[6] They point to the Deccan Plateau in India that outpoured 200,000 mi^2 of basaltic lava at the K-Pg boundary level and unknown amounts of sulfur and carbon dioxide into the atmosphere.[7] This was enough lava to cover the states of Texas and Alaska to a depth of 2,000 feet. They argue that the climatic changes from such an eruption would have wiped out many life forms.

The size of the hypothetical K-Pg asteroid may not have been as large as previously thought. Other secular authors have reduced the estimated size of the K-Pg asteroid to only 2.5 miles in diameter (as opposed to the previously reported 6 to 10 mi).[8] They used the amount of super-dense Osmium (Os) in the ocean sediments above and below the K-Pg layer to calculate the size of the impacting body. Because the percentage of Os isotopes in meteorites is known, they were able to estimate the approximate diameter of the asteroid from the distribution of Os isotopes in the sedimentary record. The lower-than-expected levels of Os they found forced their reinterpretation of a smaller asteroid.

Even the 26- to 28-million-year cyclic extinction pattern is being questioned by secular scientists. Critics point out that the presumed pattern was found as a consequence of the sampling interval. The original proponents, they argue, sampled the extinction record every 6.2 million years or so through time, using only limited intervals corresponding to geologic stages. Critics contend, therefore, that the result is close to a simple mathematical multiple of the sampling interval ($4 \times 6.2 = 24.8$), creating a false cycle to the data. They also found that the spacing between sampled intervals only allowed resolution of periods above 11 or 12.5 million years, which again, doubled, is about 25 million years. In addition, some of the calculated extinction cycles do not correlate well with the extinction events within the sediments.

A more recent and comprehensive study looked at fossil diversity through geologic time, sampling every 1 million years of the secular time scale.[9] They used the 36,000 species in the compendium compiled by J. Sepkoski, running Fourier transforms of the data, and found virtually no evidence to support a 26- to 28-million-year cycle in the data set.

Admittedly, there likely were several asteroid impacts during the Flood as part of the one-year global judgment. Creation scientists don't think these caused any extinctions, however, as God protected Noah and the animals on the ark, even from extraterrestrial impacts. But because secularists ignore the evidence of a global Flood, they will likely continue to debate the causes of the K-Pg extinction for many years.

WHAT THE ROCKS TELL US

As discussed earlier, "extinctions" seem to occur because there is a rapid change in rock type and/or depositional pattern at or near a particular rock boundary, like the K-Pg boundary. These rapid changes in rock type would seem to rapidly "finish off" a population, like the dinosaurs in the Hell Creek Formation and elsewhere. Rapid shifts in depositional patterns are an expected consequence of a flood of global proportions.

A review of 110 secular research papers on the K-Pg extinction, examining all organisms, showed that a sudden extinction was noted in only 50 percent of the papers, and that 72 percent of those sudden extinctions were associated with a corresponding rapid shift in rock type.[10] This strong correlation demonstrated that the perceived abruptness of the K-Pg extinction is enhanced by observable rock changes. This is just what you would expect as Flood sediments were being piled on top of one another, accumulating new rock types on new rock types, and new organisms appearing and disappearing rapidly. We also found that 28 percent of the extinctions occurred prior to the K-Pg boundary, 55 percent occurred at the K-Pg boundary, and 17 percent occurred just after the K-Pg boundary, in the Early Tertiary system. So it appears there is no definite boundary where all organisms, including the dinosaurs, went extinct to mark the so-called K-Pg boundary. Organisms merely stopped being deposited at those locations. It's no wonder evolutionists are so confused. It all goes back to their worldview once again.

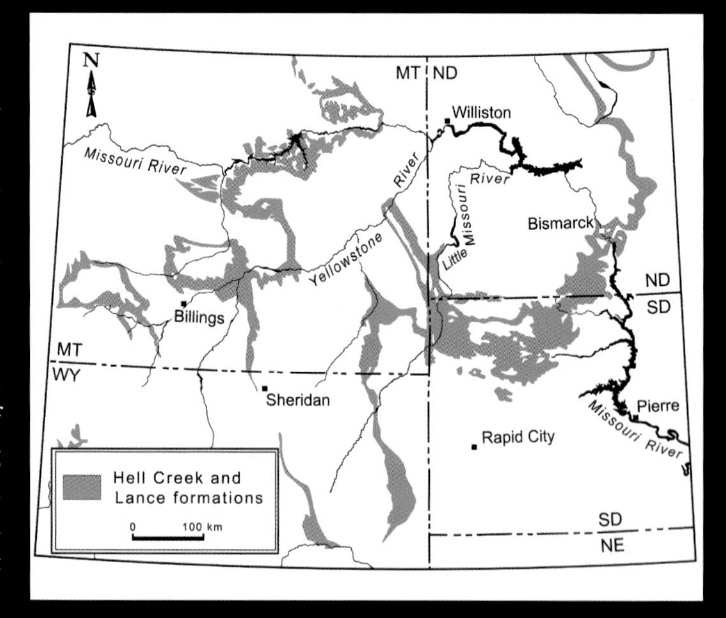

WHY DINOSAURS ARE FOUND ONLY IN LATER FLOOD STRATA

My research has involved the examination and analysis of hundreds of stratigraphic rock columns across the USA and Canada. The results have provided clear insight into depositional patterns of the Flood across North America. First, deposition of the first three megasequences (Sauk, Tippecanoe, and Kaskaskia) was thickest in the eastern half of the United States, with accumulations commonly greater than 2 miles. By contrast, the early Flood deposits across much of the West were much thinner at this juncture in the Flood, with thicknesses in many areas of less than a few 100 yards. And in many places, there appeared to be no deposition of the early megasequences at all.

Fossils of dinosaurs do not show up in the earliest Flood sediments. Instead, the earliest three megasequences are dominated by shallow marine animals, such as brachiopods, corals, trilobites, and echinoderms. There are even some fish and cephalopods mixed in these earliest sediments. Also, these first sequences do not contain large marine reptiles, such as mosasaurs and plesiosaurs and ichthyosaurs, nor are there any dinosaurs in these sediments.

The dinosaurs were able to survive through the early Flood sequences in the West, from Texas to Montana, because they were able to remain on (or swim and scramble to) the exposed areas of high ground as the Flood waters advanced. I call this high ground "dinosaur peninsula." It roughly coincides with what secular geologists call the Ancestral Rocky Mountains and the so-called Transcontinental Arch. Without getting into too much technical detail here, suffice it to say there were ample "islands" that received no deposition until the later megasequences (Absaroka and Zuni). During the early days of the Flood, dinosaurs were still able to find limited land to congregate on, even while active sedimentation was occurring nearby. This is also why we see no dinosaur footprints in the early three megasequences, and later we see many tracks very close together within the Absaroka and Zuni megasequences as if they were herded in these locations. The sites and numbers of dinosaur footprints are so plentiful along a narrow zone from Texas, up through Colorado, Wyoming, Montana, and into Alberta, that it has been called the Dinosaur Racetrack (chapter 11). There were a lot of dinosaurs packed in a limited area at this point in the Flood.

Later in the Flood, at the time of deposition of the Absaroka and the Zuni megasequences, things had changed dramatically. Movement of the tectonic plates had started to cause subduction along the West coast, and new ocean crust began to appear in the east as Pangaea began to break up. This change in tectonics, combined with increasing water levels, caused great changes in depositional style. Tsunami-like water waves washed violently across the continent from the west. It is at this time, and across the world, that we observe the largest, cross-bedded sandstone layers, some greater than 20 feet high. These remnants of huge sand waves testify to the power of the Flood during this time.

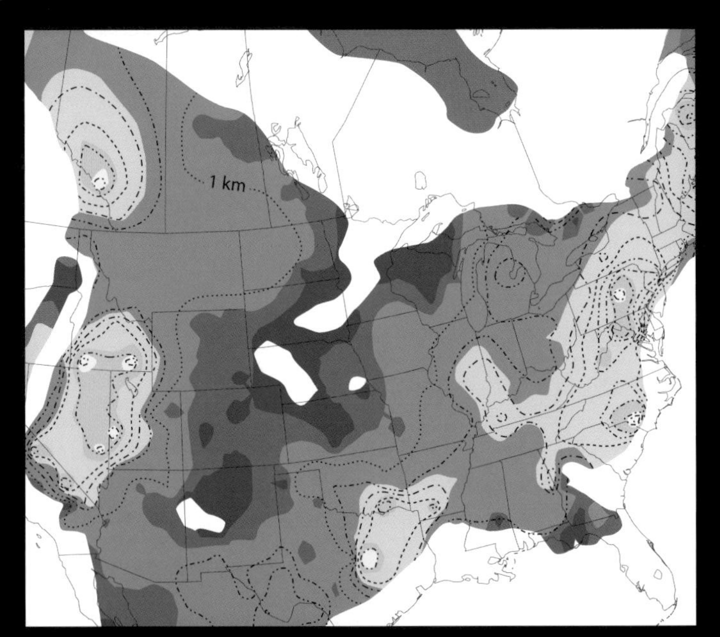

Total thickness map of the early Flood sediments, the Sauk, Tippecanoe and Kaskaskia megasequences. Note that there is a large area extending from Minnesota to New Mexico of either thin or no deposition in these first three megasequences.

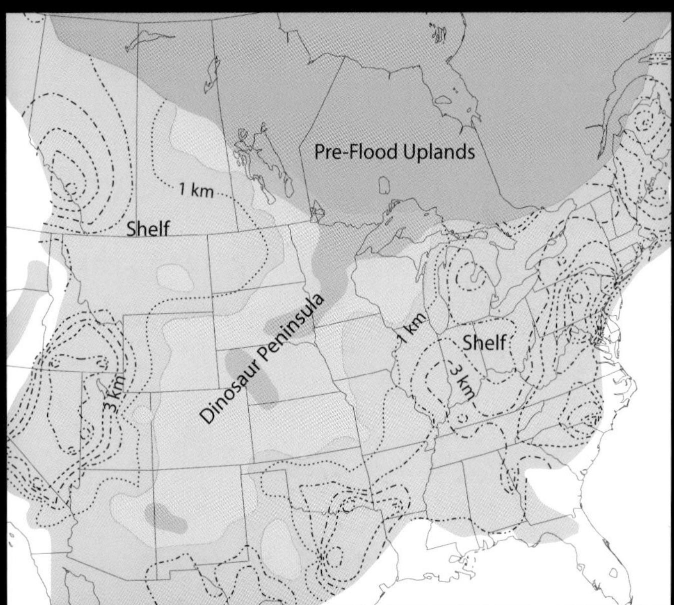

Pre-Flood geography map showing the proposed "Dinosaur Peninsula," and associated lowlands, coinciding with the thinnest early Flood sediments. Dinosaurs apparently lived at high enough elevations (possibly up to 1000 feet above sea level) to avoid becoming entombed in the earliest Flood sediments. Their mobility likely played a part in their early survival also.

Unfortunately, the secularists interpret these features as desert sand dune deposits, in spite of the obvious marine influence (such as the angle of the cross-beds), because they cannot imagine water waves of such magnitude. However, in a Flood of global proportions, huge tsunami-like waves are to be expected as the plates of the earth lurched forward at several miles per hour.

Stratigraphic column data show over two miles of sediment accumulated across the southwest part of the United States during the Absaroka megasequence, west of a line from Arkansas to Idaho. This overwhelmed and buried the Triassic and Lower Jurassic system dinosaurs which could no longer escape the Flood, as the waves and sediment engulfed them. It also began the systematic flooding of Dinosaur Peninsula from south to north as the Flood waters rose yet higher.

As deposition of the Absaroka came to a close, the Zuni megasequence followed. Here again, we observe that the thickest sediment accumulation occurred in the American West and Canada, with very little in the East. The Zuni depositional center was west of a north-south line from the Dakotas to the Texas Panhandle, where sediments thicker than two miles again are common. This additional sedimentation trapped and buried the Upper Jurassic system dinosaurs and the Cretaceous system dinosaurs in rapid succession, completing the destruction of the dinosaurs along the central and northern section of Dinosaur Peninsula, respectively.

Huge water waves at this time were, again, likely a consequence of subduction. Tsunamis were generated from movements of the seafloor along the West Coast, caused by movement of the North American continent westward at several miles per hour. As deposition of the Zuni sequence ended, nearly all, if not all, the dinosaurs were inundated, and many were entombed as future fossils. The only dinosaurs to survive were those on the ark with Noah and his family.

We also find the largest herds of dinosaurs in the Upper Cretaceous system sediments in northern Wyoming, Montana, and Alberta, Canada. It's as if the dinosaurs were running northward up the peninsula as the waters were approaching from the south, cutting off any chance of escape and herding them even closer together before their demise. In his book *Digging Dinosaurs*, John R. (Jack) Horner has reported the discovery of over 10,000 adult *Maiasaura* in one location in western Montana without finding any babies. It baffled him how all these adults could be found in a small area and yet no young were mixed in with them.[11] What could have caused this odd sorting by size? Where was the social order of the herd?

In a Flood model, this is more easily explained. These adult dinosaurs were likely running in a "stampede" away from the imminent danger of raging Flood waters that were fast approaching. They weren't thinking of their young. It was complete panic. Chaos and instinct ruled the moment. The adult *Maiasaura* were running for their lives. In the process, they probably outran their hatchlings and young juveniles, sorting themselves by size and speed.

The K-Pg boundary near Glendive, Montana. The Cretaceous Hell Creek Formation, containing many dinosaurs, is below the black, coal-rich layers.

Grand Staircase Escalante National Monument, Utah. Jurassic system rocks display vivid colors.

WHY THE DINOSAURS REALLY WENT EXTINCT

Although no one knows for sure, and exactly when post-Flood the last dinosaur died, it appears that they are extinct. There is a remote chance that a "living" fossil may have existed until recently, but there is no physical evidence of any alive today. After the Flood, the dinosaurs probably slowly died off due to atmospheric and climatic changes in the post-Flood world. As discussed in chapter 2, the pre-Flood world exhibits numerous examples of gigantism and extremely large animals and insects, supporting the interpretation of higher atmospheric oxygen levels before the Flood. The massive amounts of coal and plant fossils buried in Flood sediments also indicate there was likely more carbon dioxide to sustain a prolific ecosystem in the pre-Flood world too.

Keep in mind that there were probably no extinctions prior to the Flood. All extinctions, including the dinosaurs, likely occurred post-Flood. The small populations immediately after the Flood made extinction even more likely for animals that reproduced slowly. If dinosaurs were cold-blooded, as much evidence indicates, climatic changes would have had a devastating effect on growth and reproduction rates, and maybe even on activity levels, depending on their respective latitude and elevation.

Dinosaurs apparently lived for many centuries after the Flood, giving man the chance to see, write about (the Book of Job), and make images for us to view even today. Mankind may have even hunted and killed some dinosaurs. We know from Genesis 10:9 that Nimrod "was a mighty hunter before the LORD." He, and men like him, may have hunted and killed off some species of dinosaurs.

Extinction in the rock record is defined as the highest stratigraphic occurrence of an organism as a fossil. During the Flood, layers of animal and plant communities were rapidly piled on top of one another, from various directions and in seemingly a global order, according to the tectonic activity taking place at that moment. Viewed in this manner, these last occurrences of fossil organisms of are not extinctions at all, in terms of the secular definition. Instead, they are merely the last record of various organisms trapped by the Flood waters. Some creation scientists think the "K-Pg" was about the high point of the Flood waters when all land animals finally succumbed and drowned. That may be why it is at this point that so many organisms seemed to simultaneously go "extinct." It was a major juncture in the Flood record. Research on this is still ongoing.

POST-CRETACEOUS SYSTEM DINOSAURS

Scattered dinosaur fossils have been found in rocks that were deposited above the K-Pg boundary in several locations around the world, including China, India, Bolivia, and the USA. Dinosaur fossils found in these basal Tertiary (or Paleogene) system rocks (part of the Tejas megasequence) are usually ignored by paleontologists or explained away as reworked fossils. A reworked fossil is one that was originally deposited in one rock layer and, through uplift and erosion, was removed and transported again to another younger deposit. In effect, the fossil becomes like a cobble or piece of gravel that is transported twice, first deposited in one stratum, then eroded, transported, and redeposited.

Badlands near Drumheller, Alberta, where erosion has exposed the K-Pg boundary.

Some of these Tertiary system dinosaur fossils, however show little abrasion and wear, making reworking less likely. Many of the dinosaur fossils found at levels above the K-Pg are single bones, however, and most commonly are individual teeth, which would likely show less wear. In contrast, some of the best evidence for Tertiary system dinosaurs comes from China, where dinosaur eggshell fragments are found in Tertiary rocks many tens of feet after the K-Pg boundary deposit. Eggshell is more difficult to explain as simply reworked due to its fragile nature.

Creation scientists are not surprised by a few dinosaur fossils in Tertiary rocks, as there is no reason to suspect all dinosaurs were buried below the K-Pg boundary layer. Many also agree that some dinosaur bones were probably reworked. God tells us in Genesis 8:3 that the water receded steadily from the earth after flooding everything to a depth of at least 20 feet (Gen. 7:20). This late Flood period of rapid recession probably removed and redeposited a lot of recently deposited sediment and the bones they contained (i.e., dinosaurs). Consequently, it is likely that some of these dinosaur bones became buried in younger sedimentary deposits as part of that recession.

Alternatively, it is possible that a few dinosaurs (or their floating carcasses) survived until the earliest Tertiary rocks were deposited before being entombed in sediments by the onslaught of the Flood. Creation geologists are actively trying to determine the high water level of the Flood in the rock record.

Most evolutionists insist dinosaurs never really went extinct, they just evolved into birds. Creation geologists vehemently disagree. All features of dinosaurs make them a unique reptile group that was created on day 6 as part of the creatures that move along the ground (Gen. 1:24), whereas God specifically called every winged bird into existence on day 5 (Gen. 1:21).

THE LAZARUS EFFECT AND LIVING FOSSILS

According to secular scientists, sauropod dinosaurs (the long-necks) disappear in the rock record during the Lower Cretaceous and then reappear in the Upper Cretaceous. Secular paleontologists are baffled by this apparent extinction and sudden reappearance, supposedly 30 million years later. Ironically, they have labeled this the "Lazarus effect." In order to resolve this dilemma, some secularists have suggested that these dinosaurs evolved more than once, creating an additional discrepancy in their evolutionary theory.

In contrast, creation scientists find no discrepancy with this gap in sauropod fossil occurrence. Deposition of different ecological communities in rapid succession during the Flood would be expected to cause abrupt fossil changes and so-called "extinctions" of many organisms. A gap in sauropod deposition can be easily explained by keeping in mind that these weren't true extinctions, just a record of changing conditions during the Flood. The last fossil appearance of a type of organism in the Flood may have been days or weeks after similar fossils were deposited. It's no surprise that a few animal types were deposited more than once, with some sedimentary rocks devoid of their remains in between.

The 1994 discovery of the Wollemi pine also caused quite a stir in paleontology. This tree species was commonly thought to be extinct, as no fossils of it had been found in rocks above the Cretaceous system. The discovery was dubbed the botanical equivalent of finding a living dinosaur. Creationists realize this "living

Gingko fossil leaf

Coelacanth Fossil

Living Coelacanth

fossil" never went extinct. As the Flood reached its highest level, possibly marked by the deposition of the Cretaceous rocks, there was less opportunity to bury and preserve fossils, including these trees. Their existence today demonstrates that so-called extinction events are only the last appearance of organisms in the rock record. They do not represent real extinctions.

The 1938 discovery of the "living fossil" coelacanth in the Indian Ocean sent a shock wave through the paleontological community. This fish was likewise supposed to be extinct, as no fossils of it were observed in rocks above the Cretaceous system. Its discovery was the marine equivalent to finding a living dinosaur. Even more than that, this fish (which was thought to be a shallow-water coastal dweller) was found, as have all subsequent specimens, to reside at depths of about 500 feet below sea level. Traditional paleontology had this fish as an "intermediate" between fish and amphibians because it had "fleshy appendages." However, direct observation of living specimens and physiological analysis demonstrated this fish did not crawl on land as thought. Its limbs were designed for swimming and maneuvering at great depths. The disappearance of the coelacanth from the rock record may also coincide with the highest water level of the Flood.

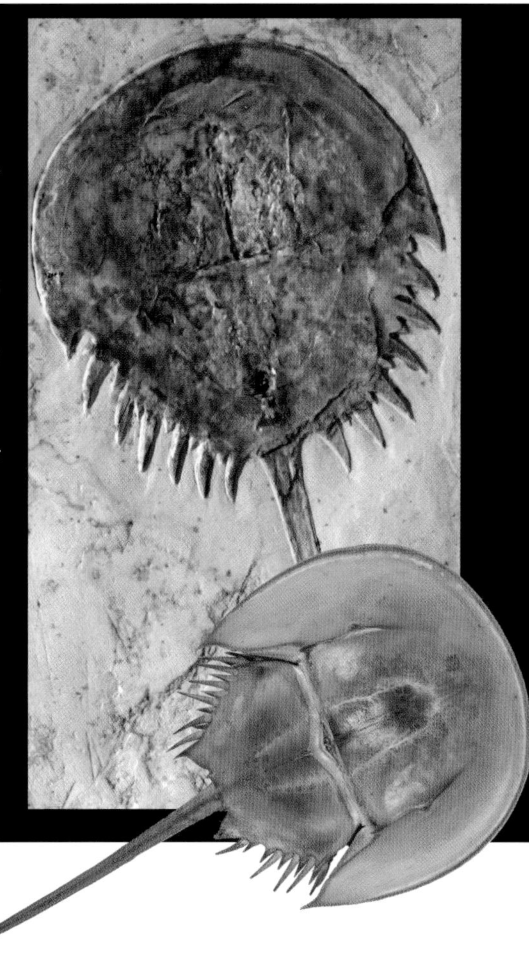

Spinophorosaurus, a fairly newly discovered sauropod dinosaur from Niger. It was unusual, as it had bony tail spikes similar to *Stegosaurus*.

CHAPTER 13: DIGGING

FINDING DINOSAUR BONES YOURSELF

You may have wondered how exactly do paleontologists find dinosaurs? Sometimes paleontologists go out prospecting for new dinosaurs, but many paleontologists rely on amateurs to find new sites. Before you begin looking for dinosaur bones, you must first gain permission from the landowner or government agency that administers the land. Permission to collect vertebrate fossils on federal land is normally granted only to universities and/or museums (agencies that will put their finds on public display). It is illegal for individuals to collect vertebrate fossils on public land. For this reason, I recommend looking only on private land, after gaining permission from the owner.

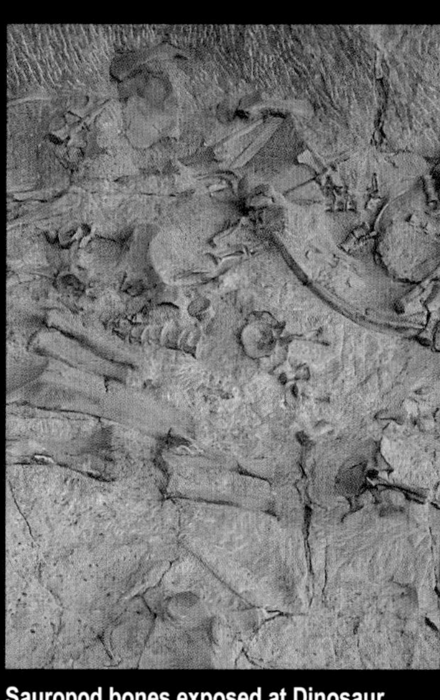
Sauropod bones exposed at Dinosaur National Monument, Utah.

Begin your search by locating rocks that will most likely contain dinosaurs. Dinosaurs are found in the so-called Mesozoic Era (late Flood) sedimentary rocks. The best sites are rocks containing a high proportion of other land-type fossils. Rocks that contain purely marine organisms are least likely to contain dinosaurs. Recall, sedimentary rocks are a record of buried ecological communities, one on top of the other. Many communities buried in the Flood waters stayed quite distinct from one another. Although there are isolated cases of dinosaur fossils found in marine deposits, having their bones washed out to sea, these occurrences are so rare it is best to avoid them and concentrate on where they are more prevalent.

It is helpful to have some basic field gear for prospecting. What if you actually find a dinosaur? I recommend carrying around the following items at least: backpack with many pockets, field notebook, canteen (2 liters plus), geologic maps, topographic maps, rock hammer, compass, waterproof pen, pencils, paintbrushes (2-inch), cloth collecting bags, ice pick or dental tools, Swiss army knife, Superglue, plastic bags, measuring tape (metric preferred), camera, and specimen labels. I know it's a lot to carry, but you may wish you had more if you find something. And keep in mind, your cell phone may not have coverage in remote areas, so it's best to take a second person along for safety.

Begin by looking at the base of hills in "badlands" where a lot of rock is exposed. You can also look along dry stream beds where they cut through rocks. These types of areas will receive all the eroded fragments of bones that wash down from the surrounding hills and cliffs. The washed-down bone fragments are called float. If you discover some bone, follow the trail up the hill or cliff face, looking for the in situ (in place) source of the fragments. Usually the trail will lead to a sandstone layer where bone may still be sticking out of the rock.

Digging for hadrosaur hatchling fossils in Montana.

Dinosaur bone fragments at the surface. 35-mm camera lens cover for scale.

Some of the tools used to excavate dinosaur bones, including old dental tools.

What does fossil bone looks like? Many people can walk right by bones and not even notice them. Lucky for us, fossil bone is often a different color than the surrounding rock. It can be black, white, or tan, depending on the mineralization that took place. Sometimes it is more shattered and crumbly than the rock. The best way to identify bone, however, is to look for the texture of bone. Bone is porous. It will absorb water easier than rock and will contain "holes" if the insides are exposed. Sometimes the outer surface will be layered.

As a final test, if you're in doubt, try placing the presumed dinosaur bone fragment up to your tongue. If it sticks, it is definitely bone. The porosity of the bone causes this "sticking." The surrounding rock will just fall to the ground if you try this test.

Brushing loose debris away from excavated dinosaur bones.

Once you've followed a bone trail up a hill and found some in situ dinosaur bones hanging out of the rocks, you can begin your excavation. First, brush away the loose dirt and debris with one of your paintbrushes. Do not use your fingers, as the pressure can cause excessive damage. Next, expose a bit more of the specimen with small hand tools, such as dental picks or ice picks. If the specimen is crumbly or soft, add a little bit of Superglue for strengthening.

The next step is to document the site with photography and record the location on an appropriate topographic map. A GPS system is handy at this step also. Small bone fragments can be collected in cloth or Ziploc bags. Make sure you label the bags with detailed information about the site. If you find more bone remains in the rock face, or a lot of bones, such as in a bone bed, you'll want to seek help from a professional. Don't worry, you will still be famous as the discoverer of something new or special, even if you don't do all the excavation work. Once a site has been located and deemed worthy of excavation, you may want to join the professional paleontology crew in digging at the site.

DIG SITE TECHNIQUES

You may want to consider digging at an established site instead of looking for your own bones. Finding new specimens can be tedious and requires a lot of patience. A commercial dig site can cost considerable amounts of money, especially to "volunteer" with the crew of a famous paleontologist. Costs of $100 to $300 per day are not unusual. There are many sites across the nation. A quick search of the web will offer plenty of choices.

If you actually go to a dig site and participate in a dinosaur dig, you will want to keep in mind some of the following techniques and dig site etiquette. Always approach a potential bone from above, removing rock overburden slowly and carefully, and in small pieces. Use your brush more than metal tools. We were always told, "A happy dig site is a clean dig site." This rule cannot be underestimated. Also, the paintbrush is often your best and most effective tool. We were also told, "To find bone you must break bone." This doesn't mean you continue to break the bones, but any new bone must first be found by breaking it free from the rocks. This also means you need to carefully examine each rock fragment, on all sides, before placing it in the discard pile. I almost threw away an *Allosaurus* tooth that way. Some bones are very small. Examine every side to every piece.

Author's daughter helping to survey in the location of a dinosaur bone.

Step one in the discovery of a *Brontosaurus* tail vertebrae.

Some sort of GPS grid system is usually utilized to document the bones found at a site, including the orientation of the bones. Many crews install pre-surveyed stakes in the ground at various locations to assist in mapping out the bones. Computer diagrams of the sites are becoming the norm. Paleontologists usually compile these in the winter months when digging is unavailable.

Once you find a new bone, you need to work some distance away from the bone with your hand tools so you don't break it again. Ultimately, you will need to leave a lot of the rock "matrix" attached to the bone so that it can be transported to the laboratory for further cleaning. Your job in the field is to minimize breakage until the bone can be properly prepared by professionals in the museum lab.

Each exposed bone should be "pedestalled" prior to removal. To pedestal a bone, you remove sediment around the bone/matrix specimen, but not underneath. Essentially you create a moat around the fossil that goes down deeper than the in situ fossil.

Step two: pedestalling the bone.

Step three: the author plastering the tail bone after covering it with foil and wet paper towels.

Waiting for the plaster to harden.

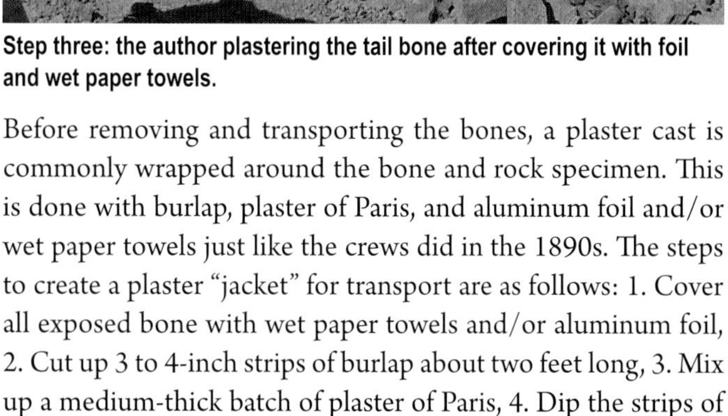

Before removing and transporting the bones, a plaster cast is commonly wrapped around the bone and rock specimen. This is done with burlap, plaster of Paris, and aluminum foil and/or wet paper towels just like the crews did in the 1890s. The steps to create a plaster "jacket" for transport are as follows: 1. Cover all exposed bone with wet paper towels and/or aluminum foil, 2. Cut up 3 to 4-inch strips of burlap about two feet long, 3. Mix up a medium-thick batch of plaster of Paris, 4. Dip the strips of burlap into the plaster and smooth on the pedestalled bone one strip at a time. The plaster cast should be about 0.5 to 2 inches thick, depending on size. Big bones may need internal wooden supports within the cast also. The cast should dry in about an hour or two.

Don't forget to carefully record the bone number and location on the plaster cast. After three or four of these, you won't remember which is which, and the lab will not know where the specimens came from.

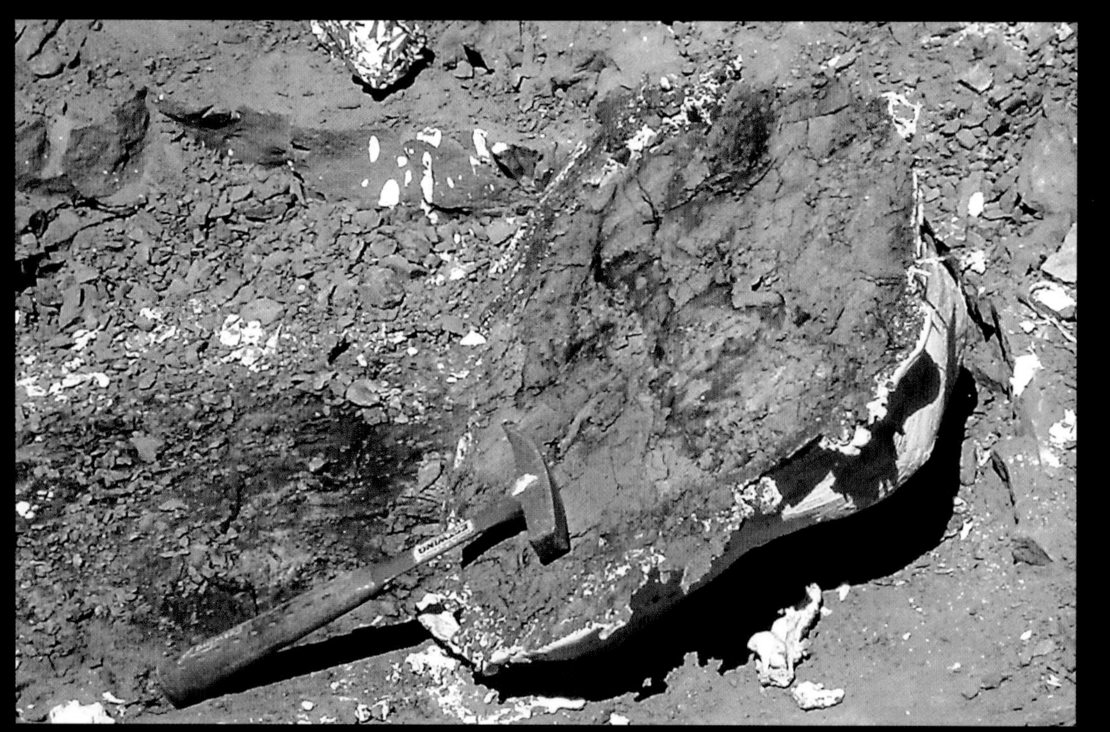

Step four: breaking the rock beneath the plastered bone. The final step may be to plaster the other side and encase the bone and rock together for shipment to the laboratory. Once there, the final excavation from the matrix material will be completed.

Finally, you need to break the bone free from the ground. This may require a strong push on small specimens or chiseling beneath larger specimens. Additional burlap and plaster is then usually placed over the exposed rock beneath the overturned specimen, resulting in a complete jacket around the fossil. The fossil is now ready for transport. Your fieldwork on that specimen is done. Now you are ready to move on to your next discovery!

Surveying in a dinosaur bone with a tape measure.

Dinosaur bones often come in a "log jam" from rapid burial and transport.

A dig site in western Montana.

Hadrosaur skull fragments on the surface. Note corrugated look to the bones.

The author's daughter (left) and son (right) helping to find dinosaur bones.

CHAPTER 14: THE REAL STORY OF THE DINOSAURS

Few things strike nearly universal interest across every age level like dinosaurs. People around the world are fascinated by these strange and marvelous creatures. Museums house thousands of skeletons and bones of these exotic animals, from the smallest, bird-sized ones to the 131-foot-long behemoths. Unfortunately, most of the museum displays, popular publications, TV shows, and movies about dinosaurs are controlled by secular scientists. They use the subject of dinosaurs to indoctrinate the masses with evolutionary dogma, starting at a young, vulnerable age. The science of dinosaurs is rarely presented in reverence to the Word of God. Although this book conflicts with the secular story of dinosaurs, it does not shy away from science. My goal in this book is to present up-to-date scientific data in a biblical context.

This chapter is both a review and a summary of what I like to call "the real story" of dinosaurs. Some of what I present is based on a lifetime of studying the geology across the United States, including my ongoing research into the megasequences. All conclusions are based on the Bible and sound science, but some topics will obviously be more speculative. I will attempt to make clear those concepts that are well-established and distinguish them from those concepts that are more nebulous.

PRE-FLOOD DINOSAURS

As discussed in chapter 1, God created dinosaurs on day 6 of creation week (Genesis 1:25). They are included in the "beasts of the earth" and in "every thing that creepeth upon the earth" (KJV).

> And God made the beast of the earth after his kind, and cattle after their kind, and every thing that creepeth upon the earth after his kind: and God saw that it was good.
>
> —Genesis 1:25; KJV

God gave every plant and herb to the dinosaurs for food (Genesis 1:30). Dinosaurs were originally created to be vegetarians, as was all flesh. It wasn't until after the Fall, after the sin of man, that a change occurred in the eating habits of some of the dinosaurs, causing them to eat flesh. Crushed bone fragments found in coprolites, teeth marks on bones, and even teeth embedded in bones of other dinosaurs support this conclusion.

For the next 1,600 years or so, the earth and mankind became increasingly evil (Genesis 6:5). The Bible tells us the wickedness of man grieved God at His heart (Genesis 6:6) "and the earth was filled with violence" (Genesis 6:11; KJV) and "all flesh had corrupted his way upon the earth" (Genesis 6:12; KJV). This corruption would have included the dinosaurs, which were actively preying on one another at this time.

Many pre-Flood dinosaurs likely lived to be hundreds of years old, as did pre-Flood humans, and consequently grew to immense sizes. The large skeletons on display at museums testify to the large sizes of these reptiles. Gigantism of many types of animals suggests a different pre-Flood atmosphere existed, with higher oxygen levels. And if cold-blooded, as substantial evidence suggests (chapter 10), the dinosaurs would have had no trouble remaining active as the pre-Flood world seems to have had a single, universally warm climate. As evidence of this activity, we see numerous broken bones and injuries in many dinosaur skeletons that testify to the violence that filled the earth. Everything from broken ribs, fractured vertebra, to infected toe bones, and often evidence of multiple bison. Sure, there were some small dinosaurs, but if you weren't big, you were likely being pursued, even by large mammals.

Next, we see God "repented" that He made man, beast, and the creeping thing (dinosaurs) (Gen. 6:7; KJV).[2] He brought on the great Flood to destroy all flesh except Noah and his family, "and of every living thing of all flesh, two of every sort shalt thou bring into the ark, to keep them alive with thee" (Gen. 6:19; KJV). Therefore, God brought two of every kind of dinosaur to the ark. There is no hint of exclusion here in the text. Dinosaurs were taken on the ark. Exactly how many we can't be certain, but likely around 60 pairs (the number of dinosaur families). The dinosaurs and all other animals were able to walk to the ark, as the majority of the pre-Flood world was likely a single, connected land mass.

How did the dinosaurs fit on the ark? God likely did not bring fully grown, adult-sized dinosaurs to the ark. So it really didn't matter that the typical dinosaur was fairly large. God likely brought young juveniles to the ark that hadn't gone through their growth spurt (i.e., teenage years in humans). In this way, the dinosaurs themselves would have been smaller, without the need to eat great quantities of food while on the ark, and yet be able to grow quickly and repopulate the earth in the immediate years after the year-long Flood ended. All dinosaurs must have eaten plants while on the ark, as originally created. There is no indication that "meat" was brought on the ark for any animal, including mammals (Gen. 6:21).

Outside the ark, pre-Flood dinosaurs seem to have lived in separate areas from humans and most of the larger mammals. They probably lived in different ecosystems and at different elevations, with the dinosaurs dominating the lower levels and the mammals dominating the uplands (more on this below). The sedimentation patterns early in the Flood suggest there were few high mountains like we have today, more likely, just broad hills and uplifted areas of higher elevation. My research suggests the pre-Flood eastern United States had the lowest elevation, with the west a bit higher (containing most of the dinosaur fossils), and the northern section of the continent was the highest land of all.

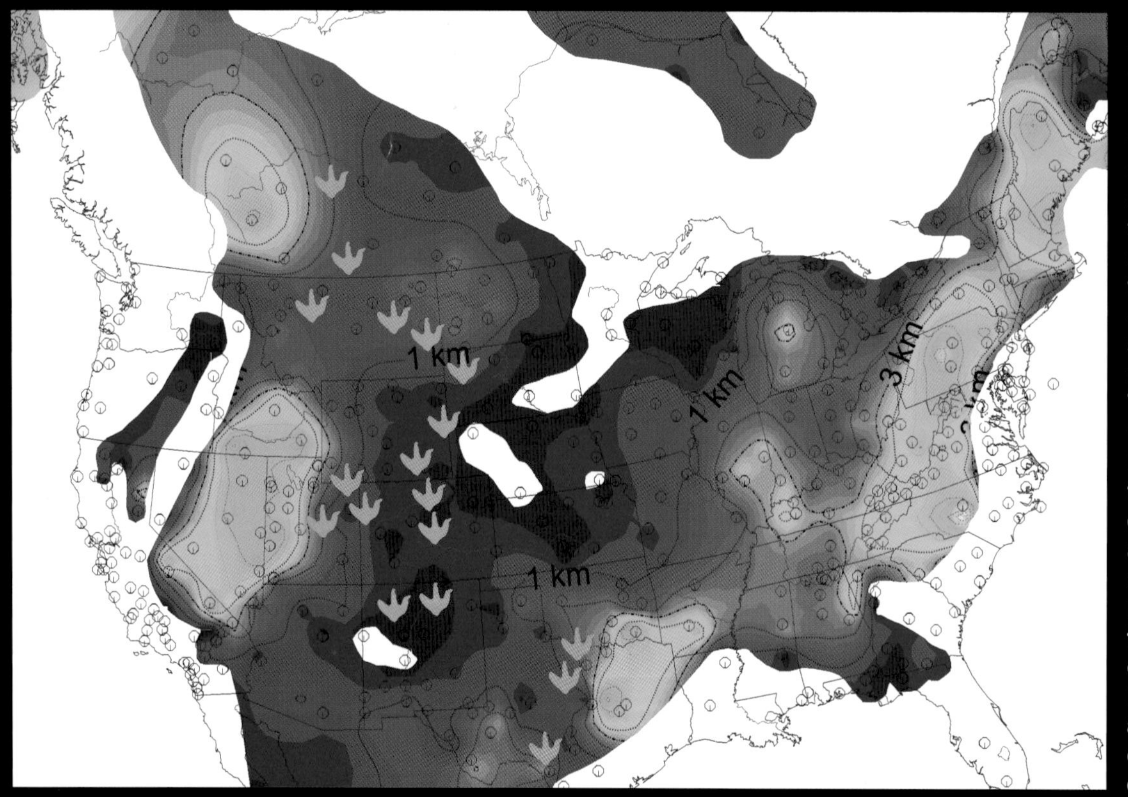

Map of the combined thickness of the Sauk, Tippecanoe, and Kaskaskia megasequences showing the sites of major dinosaur discoveries and quarries across the West. Note the coincidence of the dinosaur sites with the thinnest areas (darkest colors) of early Flood sedimentation (white represents zero thickness). This thin trend, termed "dinosaur peninsula," likely allowed dinosaurs to survive through the early part of the Flood, as sedimentation was minimal. Some of the largest dinosaur herds are found in the uppermost dinosaur-bearing strata in Montana and Alberta and appear to have been swept off the peninsula by strong currents.

THE FLOOD BEGINS

As the Flood began, the Bible tells us "all the fountains of the great deep broken up, and the windows of heaven were opened. And the rain was upon the earth forty days and forty nights" (Gen. 7:11–12; KJV). It is likely that this intense rainfall was sustained by the outpouring of tremendous quantities of extremely hot, hydrothermal waters from the upper mantle as part of the fountains of the great deep. Rift zones ran the lengths of today's major continents, pouring out lava and water simultaneously across the globe. Rapid evaporation would be expected from waters of such extreme temperatures, flashing to steam immediately as it reached the surface, as do geysers today.

In addition to the eruptions of tremendous volumes of lava and water, the continental crust was actively being pulled downward into subduction zones, warping the crust, as the plates broke and began to move. Simultaneously, rapid withdrawal of water and magma from the "fountains" created broad, collapsed basins that filled with early Flood sediments adjacent to the rift zones. These basins became depositional centers during the early Flood, sinking rapidly and collecting thousands of feet of sediment in their midst. These so-called intracratonic basins formed in the interior of many continents, during deposition of the Sauk, Tippecanoe, and Kaskaskia megasequences. If dinosaurs did live in these areas in the pre-Flood world, they either migrated or were washed away to other locations. As a result, all of the major dinosaur quarries in the United States are located on the edges or outside these major depositional basins, including the Williston, Permian, Hudson Bay, Illinois, and Michigan Basins.

Megasequences are defined as discrete packages of sedimentary rock bounded top and bottom by erosional surfaces, with coarse sandstones layers at the bottom (deposited first) followed by shale and then limestone at the top (deposited last). The corresponding size of the sedimentary particles is also thought to decrease upward in each package of rock, from sands and conglomerates at the base to clay-sized particles and fine carbonate above. The basal sandstone layers are thought to represent the shallowest sea level or a near-shore environment, the shale layers a little deeper water environment, and the limestone deposits the deepest water environment in each sequence. By tracking these changes in rock types, geologists are able to define each megasequence.

According to secular geologists, subsequent megasequences are supposed to demonstrate this same pattern of sandstone to shale to limestone as sea levels repetitively rose and fell, again and again, over millions of years, flooding the continent up to six separate times. The upper erosional boundaries in each megasequence are believed to have been created as the next younger megasequence eroded the top of the earlier sequence as it advanced landward. Secular scientists then use these megasequences to infer past environments, and of course, as an argument in support of deep time.

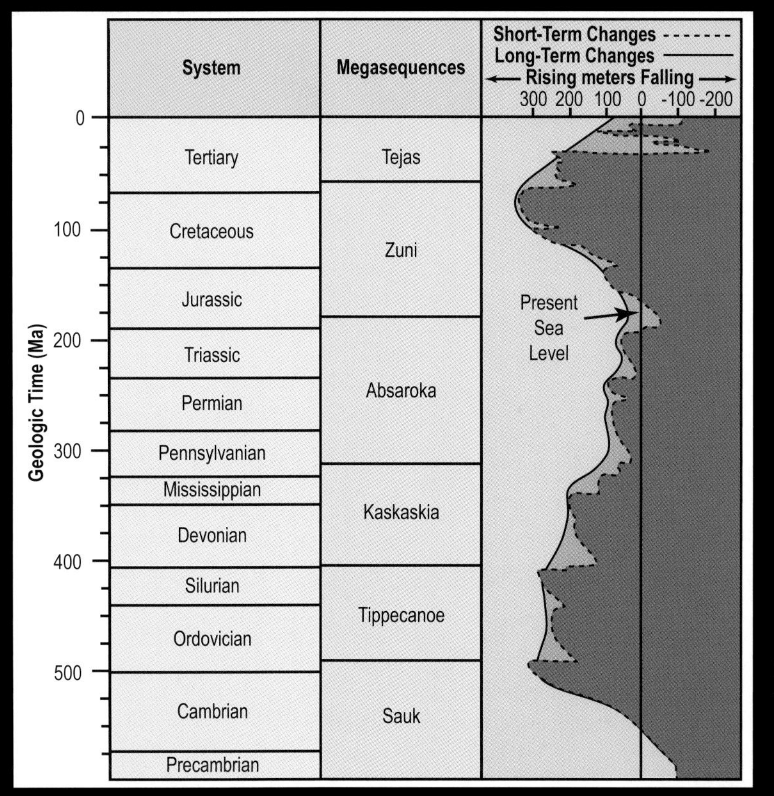

Creation geologists see these megasequences as catastrophic deposits left behind by six major advances of the Flood waters onto the continents. Each sequence seems to represent a distinct flooding episode, all within the year of the Flood, and all part of the one, catastrophic Flood event. The Bible tells us the waters of the Flood "assuaged" or oscillated. These individual advances, and subsequent minor "retreating" of the waters in between advances, resulted in the stacked megasequences we observe today. In this way, stacked megasequences are more easily explained by the sequential action of chaotic and rapidly advancing waters of the Flood.

Secular geologic time scale, sea level curve, and megasequences. The megasequences appear to record major shifts in the ocean level as tsunami-like waves came and went across the continent during the year-long Flood. Caution must be exercised to not take the exact levels of the ocean literally, as the sea level curve was determined based on uniformitarian assumptions. Diagram modified from the original provided by Dr. Andrew Snelling, *Grappling with the Chronology of the Genesis Flood* (Green Forest, AR: Master Books, 2014).

Examples of K-T (K-Pg) boundary rocks from around the world.

Colorado

Hell Creek Formation in Montana

Alberta, Canada

Netherlands

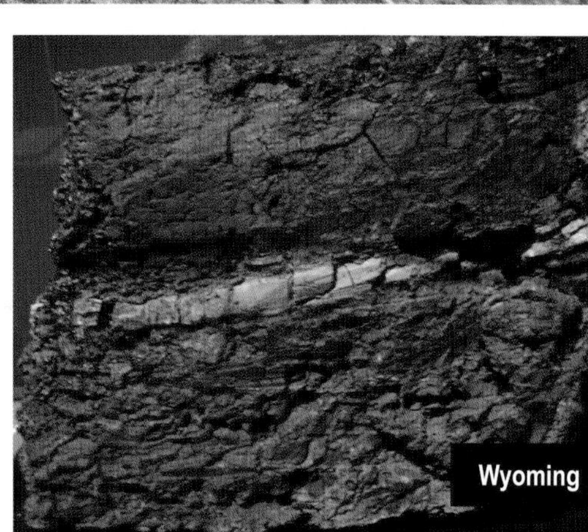

Wyoming

HOW DINOSAURS SURVIVED THE EARLY FLOOD ACTIVITY

Across much of the American West, and away from the deep intracratonic basins, the dinosaurs were able to avoid becoming entombed in the early Flood sediments by finding remnants of dry land. Along this "Dinosaur Peninsula" only limited deposition was taking place early in the Flood. My thickness map of the combined Sauk though Kaskaskia megasequences shows a marked thin zone running northeast-southwest across the American West. This "peninsula" coincides with the so-called Ancestral Rocky Mountains and the Transcontinental Arch described in many historical geology textbooks. This was the area where dinosaurs were able to find enough "dry" land to nest, lay eggs, and leave tremendous amounts of footprints along a pathway from Texas through Colorado, Wyoming, and into Alberta, Canada. This temporary refuge for the dinosaurs is also why we don't see footprints of dinosaurs in rocks deposited early in the Flood (Sauk through Kaskaskia megasequences).

Another reason dinosaurs were able to survive up to this point in the Flood was due to the lower volumes of sediment accumulation in the first three megasequences, compared to the later megasequences. In fact, our calculations based on a 3-D model of sediment volume across the USA, show only about one-third of the Flood sediments were deposited in the early stages of the Flood (the first three megasequences), and most of this was east of the modern Mississippi River. It wasn't until the latter part of the Flood when the sedimentation volume greatly increased (the last three megasequences), especially in the West, that the dinosaurs were trapped in mud and sand and buried.

Tracks in rock at Dinosaur Ridge, Morrison Fossil Area, Jefferson County, Colorado.

DINOSAURS BECOME BURIED IN THE LATER FLOOD ROCKS

After the first three megasequences, however, the Flood seems to have increased its fury and energy level, depositing nearly two-thirds of its sediment load on to the North American continent. "And the waters prevailed, and were increased greatly upon the earth; and the ark went upon the face of the waters. And the waters prevailed exceedingly upon the earth; and all the high hills, that were under the whole heaven, were covered" (Gen. 7:18–19; KJV).

It was at this point in the Flood, during deposition of the Absaroka and Zuni megasequences, that the plates seem to have undergone their most extensive episode of movement and an entirely new ocean crust began to form. The new ocean crust was hotter and less dense than the old oceanic crust. In effect, this raised the level of the ocean floor, especially near the ocean ridges, similar to the way hot air causes a balloon to rise.[3] Some creation scientists have estimated this action alone could have raised global sea level by as much as one mile, greatly helping to flood the land masses.[4]

It was during this time that subduction commenced along the west coast of North America (and other locations around the Pacific), pulling down part of the crust of North America along with it. Huge tsunami-like sand waves would have traversed from west to east across much of the continent. We find evidence of these large sand waves in the thick, cross-bedded sandstones of the Permian and Jurassic system rocks (Absaroka megasequence).

The dinosaurs were quickly inundated by the accumulating wedge of thousands of feet of sediment dumped on the Western states. As the Flood waters and sediments approached from the west, Dinosaur Peninsula was systematically buried from south to north. This stage of the Flood also washed large marine reptiles onto the continent and deposited them in the same rock layers as the dinosaurs, in many cases, mixing in sharks and fishes and clams in the same rocks as the dinosaurs, as we see in the Hell Creek Formation in eastern Montana.

Makoshika State Park badlands formations

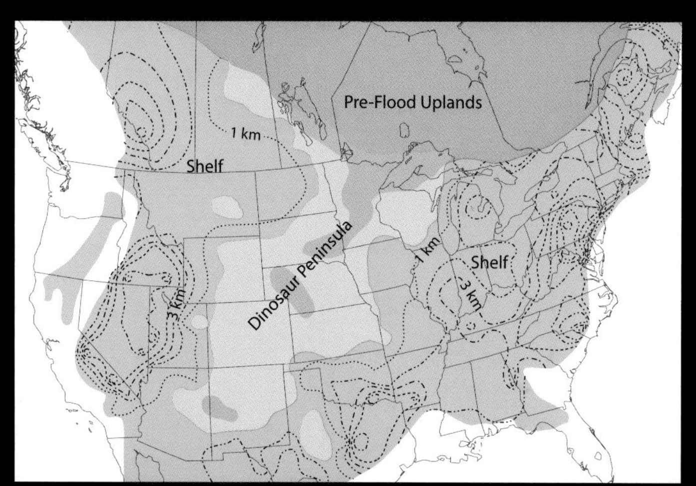

Map of pre-Flood geography showing Dinosaur Peninsula and the higher Uplands to the north in what is now Canada. This interpretation is derived from the thickness relationships observed in the earliest Flood megasequences, the Sauk, Tippecanoe, and Kaskaskia.

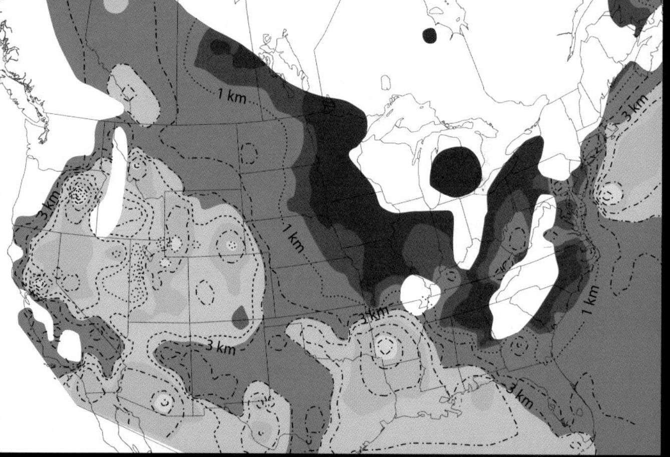

Total thickness map of the combined Absaroka and Zuni megasequences, the episode when all dinosaurs were buried by Flood sediments. Note how these sediments are thickest in the southwest and west. Dinosaur Peninsula was inundated from south to north.

> All in whose nostrils was the breath of life, of all that was in the dry land, died. . . . And the waters prevailed upon the earth an hundred and fifty days
>
> —Genesis 7:22–24; KJV

> And God made a wind to pass over the earth, and the waters assuaged; the fountains also of the deep and the windows of heaven were stopped, and the rain from heaven was restrained; and the waters returned from off the earth continually: and after the end of the hundred and fifty days the waters were abated —Genesis 8:1–3; KJV

We find few dinosaur fossils above the top of the Zuni megasequence. This level roughly corresponds to the K-T or K-Pg boundary level. The few that are found above this level may be mostly reworked material from subsequent uplift and redeposition. The last megasequence, the Tejas, is mostly filled with a combination of mammal and marine fossils.

Creation scientists differ in their interpretations at this point. Some say the Tejas megasequence is post-Flood deposits, and others think it is late Flood deposits. I favor the latter, as the tectonic plates were still moving rapidly, creating vast amounts of ocean crust during the deposition of this last megasequence. Rapid subduction and catastrophic seafloor spreading would surely have caused tsunami-like waves, tremendous earthquakes, and large-scale sedimentation to occur on a global scale. It would have been extremely difficult for air-breathing animals to have survived during this catastrophic activity, aside from those on the ark. In addition, tremendous coal beds, some approaching 200 feet thick, were deposited in eastern Wyoming as part of the Tejas megasequence. It is hard to imagine how these coal beds could accumulate without Flood waters to gather the necessary organic debris. More research is needed to resolve these differences of opinion. However, everyone agrees that any dinosaurs not on the ark at this point were likely dead.

In my opinion, most of the sediment that accumulated as part of the Tejas megasequence was associated with the receding water phase of the Flood. Approximately 80 percent of the world's mountains formed at this time, including the Rockies, as the continents began to adjust to changes in crustal thickness and mantle movements beneath (reeling from catastrophic subduction). Many areas began to uplift or rebound, while adjacent areas sank to great depths. Much of the Tejas was dominated by intermountain deposition, as the tops of the exposed mountains were stripped of their freshly deposited sediments by receding Flood water. Across the West, we observe tens of thousands of feet of Tejas equivalent sediment dumped in basins between mountain ranges. As the Flood water receded further, tens of thousands of feet of additional Tejas sediments were washed offshore into the Gulf of Mexico and along the Atlantic seaboard.

Thickness map of the Tejas megasequence. The thickest deposits are in the West and offshore, including a tremendous amount in the Gulf of Mexico. This may be the receding phase of the Flood.

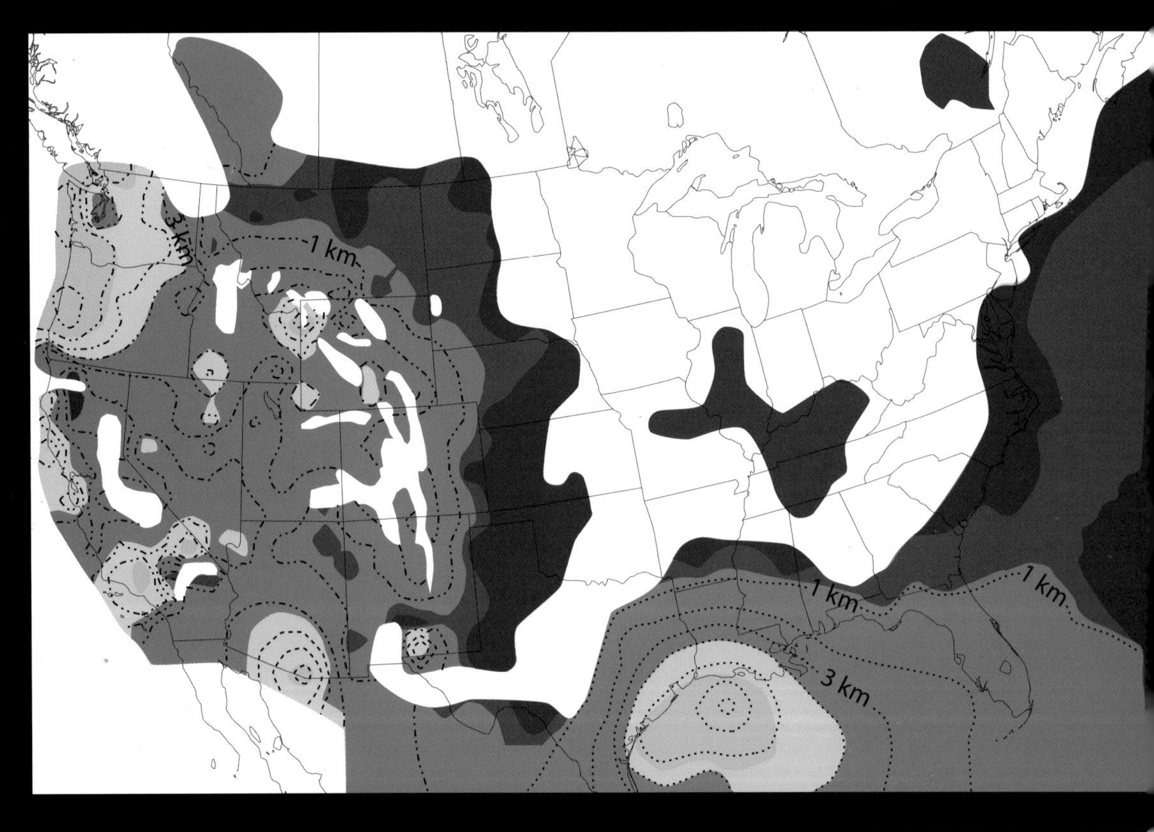

HUMANS AND MAMMAL FOSSILS

Unfortunately, there are no known, undisputed examples of human fossils mixed with dinosaur fossils. We also see few large mammals mixed in the same rocks as the dinosaur fossils. The lack of fossils also means a lack of concrete evidence. However, we do have some indirect evidence available. We can use the stratigraphic column data to infer pre-Flood highs by locating the areas of thinnest sediment accumulations.

I suspect humans and many large mammals survived until later in the Flood because they lived in, or migrated to, pre-Flood areas of highest elevation. Prior to the great Flood, the mammals in North America and the humans may have lived in what is now called the Canadian Shield. This is an area of exposed, pre-Flood crystalline rock today, largely devoid of any Flood sediments. The stratigraphic column data show all the megasequences thinning as you approach the Canadian Shield in every direction, indicating it was high ground throughout the deposition of the Flood rocks. If the dinosaurs in North America lived on a lower elevation "Dinosaur Peninsula," where most of their fossils are found, they may have been effectively separated from most of the larger mammals and humans who were living on the higher ground to the north.

Pre-Flood elevation differences may also explain the contrasting plant types that are observed in the sediments. The area of highest ground, where mammals likely dominated, was probably occupied by mostly flowering plants, as we see fossil mammals and angiosperms commonly mixed together in the same layers. In contrast, the lower elevations were dominated by nonflowering plants like ferns and conifers. Fossilized dinosaur stomach contents and fossil plants found with dinosaur remains are mostly nonflowering plants, again supporting these suggested differences in plant communities.

In addition to living in a higher elevation ecosystem, the superior intelligence of mammals and humans may have influenced their survival, helping most of them to escape entombment in the early megasequences. If the Flood waters reached their maximum height at the end of the Zuni megasequence, as has been suggested,[5] many of the large mammals and humans may have already been dead and floating in the water during the latter stages of the Flood. Also, many of their bones could have become broken and obliterated by surface wave action, destroying any potential fossils. Regardless, we do find many mammals deposited within the Tejas megasequence as the Flood waters receded. They became trapped and buried within the intermountain basins as the waters swirled about, draining off the continent.

Did humans escape burial only to rot at the surface for months as the water receded, eliminating their chance to become fossils? This may have been part of the reason Noah had to wait for seven months after the ark grounded on the mountains of Ararat. Again, further research is necessary to answer this question.

After residing on the ark for a total of 371 days, God told Noah and his family to go forth. He also commanded Noah to "bring forth with thee every living thing that is with thee, of all flesh, both of fowl, and of cattle, and of every creeping thing that creepeth upon the earth; that they may breed abundantly in the earth, and be fruitful, and multiply upon the earth"(Gen. 8:17; KJV). The numerous pairs of dinosaur kinds went forth at this point. I suspect there was a significant amount of time allotted (at least one reproductive season) before dinosaurs began eating one another again or other animals. The biblical text does not specify, but because there were only two of the unclean kinds, the death of either partner was critical to survival and would have conflicted with the command to "be fruitful, and multiply."

DINOSAUR EXTINCTION AFTER THE FLOOD

Dinosaurs apparently lived for centuries after the Flood, giving man the chance to record descriptions of what seem to be dinosaurs in writings, carvings, and paintings. Humans may have even hunted and killed some dinosaurs. We know from Genesis 10:9 that Nimrod "was a mighty hunter before the LORD." He, and men like him, may have hunted and killed off some species of dinosaurs.

However, I don't think humans were the primary cause of dinosaur extinction. They may have contributed to their demise, but it's probably more related to post-Flood climate change. The pre-Flood world seems to have had a completely different climate compared to the post-Flood world (chapter 13). The Bible is mostly silent on this too, but the fossils and rocks laid down in the Flood imply an antediluvian world with a universally warm climate. As discussed earlier, it is likely that there was more oxygen and carbon dioxide in the pre-Flood atmosphere. A drop in atmospheric oxygen level, combined with a more seasonal and cooler global climate, would be troublesome for large, cold-blooded organisms, as I suspect dinosaurs were. This new post-Flood climate would have limited their range on the earth, and the more exaggerated seasons would have affected their activity levels. If an animal couldn't breathe as efficiently, grow as fast, reproduce as quickly, and was unable to sustain activity levels, it would have been difficult surviving in the long term.

God's marvelous dinosaurs are found only as fossils today in the Flood deposits. But we are learning more and more about their diversity and complexity every year. The recent and repeated discoveries of preserved dinosaur soft tissue confirm that they are not millions of years old, only thousands, like the Bible tells us in Genesis. Their story is not an ancient story, but one that clearly shows a relevance to the world today.

> But as the days of Noah were, so shall also the coming of the Son of man be. For as in the days that were before the flood they were eating and drinking, marrying and giving in marriage, until the day that Noah entered into the ark. And knew not until the flood came, and took them all away; so shall also the coming of the Son of man be.
>
> —Matthew 24:37–39; KJV

God judged the world once about 4,500 years ago with a global Flood that Jesus spoke of as historical fact. The evidence is found in the rocks and fossils all over the world. It is clearly seen (Rom. 1:18–22). And He will come again, to judge the living and the dead. Are you ready to meet your Creator?

APPENDIX: DETERMINING THE WEIGHT OF A DINOSAUR USING SCALE MODELS

The scale model method starts with a solid, waterproof, plastic scale model of a dinosaur. It is best to use scientifically constructed models from quality institutions such as the Carnegie Museum or the British Museum for accuracy. Other equipment that is suggested is a metric ruler, a calculator, and a large graduated cylinder. Metric works best as one milliliter of water equals one cubic centimeter of volume.

Step 1	If you don't know the scale of your model, you will need to measure the length in meters (100 cm = 1 m; 1 cm = 10 mm). Most models are 1/40th or 1/25th scale. You will also need to know the actual length of the "living" dinosaur. This is sometimes written on the underside of the model. Books or a web source can also serve as a reference here. Finally, to determine your model's scale, divide the actual "living" dinosaur length in meters by the model length. For example, if the resulting number is 50, then the scale of your model 1/50th.
Step 2	Fill your graduated cylinder deep enough so that you can submerge your dinosaur model completely, but not so deep that you exceed the numbers on the side. Save a little room for the displacement of your model. Next, you need to record how much water is displaced after you submerge your model. This is a bit tricky, as "solid" plastic is somewhat buoyant. Your fingers shouldn't be submerged or you will add extra volume and reduce the accuracy of your weight estimate. Try using the tips of two pens if you are having trouble. Record your starting water volume and your "submerged" water volume. Subtract the two values to determine the volume of your model in milliliters (1 ounce = 29.6 milliliters).
Step 3	Cube the scale of your model to determine its cubic scale. If your model is 1/40th scale, its cubic scale is 40 x 40 x 40 = 64,000. This means that the model is 1/64,000th the volume of the "living" dinosaur. Next, multiply the cubic scale by the volume of your model determined in Step 2 (64,000 x your model volume in ml). You now have the "scaled up" or "living" dinosaur volume in milliliters.
Step 4	Divide the volume in milliliters by 1,000. This converts the "living" dinosaur volume to liters. Remember, one liter of fresh water weighs one kilogram. This is the beauty of the metric system.
Step 5	Here we run into a slight difference of opinion. Some suggest using a density of 1,000 kilograms per cubic meter (1.0 kg/liter), based on research by Hugh Cott, a British zoologist who worked with Nile crocodiles. Edwin Colbert and Spencer Lucas, both U.S. paleontologists, however, recommended using a density of 900 kilograms per cubic meter (0.9 kg/liter). Since no one is really sure, it's easier to use 1.0 kg/liter, allowing direct conversion from liters to kilograms. The final step then is to multiply your dinosaur volume in liters by 1.0 kilograms/liter, converting the weight of the "living" dinosaur to kilograms. A metric ton can be determined by dividing your answer in kilograms by 1,000.

You may want to convert this weight to pounds by multiplying the dinosaur weight in kilograms by 2.2 (1 kg = 2.2 lbs). To obtain the dinosaur weight in U.S. tons, simply divide the weight in pounds by 2,000.

ABOUT THE AUTHOR

Timothy Clarey has a Ph.D. in geology, two master's degrees and a B.S. in geology, summa cum laude, graduating first in his class at every level. He has authored numerous, peer-reviewed publications on the geology of the Rocky Mountain region and on the subject of dinosaurs. He worked nearly a decade in the oil and gas industry, and served as a professor at a public college for 17 years, acquiring many accolades for his teaching ability and research. He participated in numerous dinosaur digs from Wyoming to Montana and frequently took college students with him to dig sites. Dr. Clarey created and taught an undergraduate dinosaur class, writing his own college laboratory book on dinosaurs in the process. This is his second book to share his love of dinosaurs within a biblical context since joining the research staff at the Institute for Creation Research in Dallas, Texas. This book was written for the glory of the Lord, and to help gird our waists with the truth of Jesus and His creation (Eph. 6:14). He acknowledges the loving support of his wife Renee,' their children, Ryan, Ashley, Hailey and Erin, and grandchildren, his parents, Betty and Harlan, and her parents, Diania and Jim.

> For by grace you have been saved through faith, and that not of yourselves, it is the gift of God, not of works, lest anyone should boast
>
> —Ephesians 2:8–9

Skull of the sauropod *Diplodocus carnegii*, the skeleton is a copy of the original CM 84 type specimen, but the skull is a cast of USNM 2673, which is now *Galeamopus*. Late Jurassic, Morrison Formation (USA). Museo Geologico Giovanni Capellini, Bologna (Italy).

ENDNOTES:

CHAPTER 1

1 S.J. Nesbitt, P.M. Barrett, S. Werning, C.A. Sidor, and A.J. Charig, "The Oldest Dinosaur? A Middle Triassic Dinosauriform from Tanzania," *Biology Letters*, 2012, doi.org/10.1098/rsbl.2012.0949.

2 S.L. Brusatte, G. Niedzwiedzki, and R.J. Butler, "Footprints Pull Origin and Diversification of Dinosaur Stem Lineage Deep into Early Triassic," *Proceedings of The Royal Society*, 2010, B. doi: 10.1098/rspb.2010.1746.

3 M.H. Schweitzer, J.L. Wittmeyer, J.R. Horner, and J.K. Toporski, "Soft Tissue Vessels and Cellular Preservation in *Tyrannosaurus rex*," *Science*, 2005, 307.5717: 1952–1955.

4 The Research Council of Norway, "The World's Deepest Dinosaur Finding — 2256 Metres Below The Seabed," *ScienceDaily*, April 25, 2006, <www.sciencedaily.com/releases/2006/04/060425091449.htm>.

5 J.D. Morris, "How Could Noah and His Family Care for the Many Animals on Board the Ark?" *Acts & Facts*, 2005, 34 (8).

6 J. Woodmorappe, 1996. *Noah's Ark: A Feasibility Study* (Dallas, TX: Institute for Creation Research, 1996), p. 11.

7 R.B.J. Benson, N.E. Campione, M.T. Carrano, P.D. Mannion, C. Sullivan, P. Upchurch, and D.C. Evans, "Rates of Dinosaur Body Mass Evolution Indicate 170 Million Years of Sustained Ecological Innovation on the Avian Stem Lineage," *PLOS Biology*, 2014, 12.5. e1001853.doi: 10.1371/journal.pbio.1001853; T.L. Clarey and J.P. Tomkins, "Determining Average Dinosaur Size Using the Most Recent Comprehensive Body Mass Data Set," *Answers Research Journal* 2015, 8: 85–91.

8 G.M. Erickson, K.C. Rogers, and S.A. Yerby, "Dinosaurian Growth Patterns and Rapid Avian Growth Rates," *Nature* 2001, 412: 429–432.

9 Ibid.

10 J.B. Scannella and J.R. Horner, "*Torosaurus* Marsh, 1891, Is *Triceratops* Marsh, 1889 (Ceratopsidae: Chasmosaurinae): Synonymy through Ontogeny," *Journal of Vertebrate Paleontology* 2010, 30.4: 1157–1168.

11 Ernst Mayr, *Systematics and the Origin of Species* (New York: Columbia University Press, 1942), p. 120.

12 M. Balter, "Pint-sized Predator Rattles the Dinosaur Family Tree," *Science* 2011, 331.6014:134.

CHAPTER 2

1 A. Feduccia, T. Lingham-Soliar, and J.R. Hinchcliffe, "Do Feathered Dinosaurs Exist? Testing the Hypothesis on Neontological and Paleontological Evidence," *Journal of Morphology* 2005, 266: 125–166. Doi: 10.1002/jmor.10382.

2 R.L. Larson, "Geological Consequences of Superplumes," *Geology* 1991, 19: 963–966.

3 S.A. Austin, J.R. Baumgardner, D.R. Humphreys, A.A. Snelling, L. Vardiman, and K.P. Wise, "Catastrophic Plate Tectonics: A Global Flood Model of Earth History," in R.E. Walsh, ed., *Proceedings of the third International Conference on Creationism, Technical Symposium Sessions* (Pittsburgh, PA: Creation Science Fellowship, 1994), p. 609–621.

4 Ibid.

5 T.L. Clarey, "Advocates for Cold-blooded Dinosaurs: The New Generation of Heretics," *GSA Today* 2007, 17.1: 45-46. Doi: 10.1130/GSAT01701GW.1.

6 V. Prasad, C.A.E. Stromberg, H. Alimohammadian, and A. Sahni, "Dinosaur Coprolites and the Early Evolution of Grasses and Grazers," *Science* 2005, 310.5751: 1177-1180. Doi: 10.1126/science.1118806.

7 P.A. Hochuli and S. Feist-Burkhardt, "Angiosperm-like Pollen and *Afropollis* from the Middle Triassic (Anisian) of the Germanic Basin (Northern Switzerland)," *Frontiers in Plant Science* 2013, 4.344. doi: 10.3389/fpls.2013.00344.

8 Robert T. Bakker, *The Dinosaur Heresies: New Theories Unlocking the Mystery of the Dinosaurs and Their Extinction* (New York: Morrow, c1986), p. 184, 197.

9 Prasad, Stromberg, Alimohammadian, and Sahni, "Dinosaur Coprolites and the Early Evolution of Grasses and Grazers."

10 John R. Horner and James Gorman, *Digging Dinosaurs* (New York: Workman Pub., c1988).

11 S.T. Grimes, F. Brock, D. Rickard, K.L. Davies, D. Edwards, D.E.G. Briggs, and R.J. Parkes, "Understanding Fossilization: Experimental Pyritization of Plants," *Geology* 2001, 29.2: 123–126, Doi: 10.1130/0091-7613(2001)029<0123:UFEPOP>2.0.CO;2.

12 B. Thomas, "A Review of Original Tissue Fossils and Their Age Implications," in M. Horstemeyer, ed., *Proceedings of the Seventh International Conference on Creationism* (Pittsburgh, PA: Creation Science Fellowship, 2013).

13 J.D. Archibald, *Dinosaur Extinction and the End of an Era* (New York, NY: Columbia University Press, 1996).

CHAPTER 3

1 D. Castelvecchi, "Half-life (More or Less)," *Science News*, November 22, 2008, 174.11:20. www.sciencenews.org/view/feature/id/38341.

2 D. DeYoung, *Thousands . . . Not Billions* (Green Forest, AR: Master Books, 2005).

3 M.H. Schweitzer, J.L. Wittmeyer, J.R. Horner, and J.K. Toporski, "Soft Tissue Vessels and Cellular Preservation in *Tyrannosaurus rex*," *Science* 2005, 307.5717: 1952–1955.

4 M.H. Schweitzer, Z. Suo, R. Avci, M.A. Allen, F.T. Arce, and J.R. Horner, "Analyses of Soft Tissue from *Tyrannosaurus rex* Suggest the Presence of Protein," *Science* 2007, 316.5822: 277–280; J.M. Asara, M.H. Schweitzer, L.M. Freimark, M. Phillips, and L.C. Cantley, "Protein Sequences from Mastodon and *Tyrannosaurus rex* Revealed by Mass Spectrometry," *Science* 2007, 316.5822: 280–285.

5 R.R. Reisz, T.D. Huang, E.M. Roberts, S.R. Peng, C. Sullivan, K. Stein, A.R.H. LeBlanc, D.B. Shien, R.S. Chang, C.C. Chiang, C. Yang, and S. Zhong, "Embryology of Early Jurassic Dinosaur from China with Evidence of Preserved Organic Remains," *Nature* 2013, 496.7444: 210–214.

6 M.H. Schweitzer, W. Zheng, T.P. Cleland, M.B. Goodwin, E. Boatman, E. Thiel, M.A. Marcus, and S.C. Fakra, "A Role for Iron and Oxygen Chemistry in Preserving Soft Tissues, Cells and Molecules from Deep Time," *Proceedings of the Royal Society B*, 2013, 281:20132741.

7 J.M. Demise and E. Boudreaux, "Dinosaur Peptide Preservation and Degradation," *Creation Research Society Quarterly*, 2015, 51:268-285 and J. Prousek, "Fenton Chemistry in Biology and Medicine," *Pure & Applied Chemistry*, 2007, 79(12): 2325-2338.

8 B. Thomas, "How Realistic Was *Jurassic World*'s Science?" *Acts & Facts*, 2015, 44(9): 20.

9 M.F. Moczydlowska, F. Westall, and F. Foucher, "Microstructure and Biogeochemistry of the Organically Preserved Ediacaran Metazoan," Sabellidites. *Journal of Paleontology* 2014, 88.2: 224–239. Doi: 10.1666/13-003.

10 J. Hebert, "Rethinking Carbon-14 Dating: What Does It Really Tell Us about the Age of the Earth?" *Acts & Facts* 2013, 42.4: 12–14.

CHAPTER 5

1 N. Ibrahim, P.C. Sereno, C.D. Sasso, S. Maganuco, M. Fabbri, D.M. Martill, S. Zouhri, N. Myhrvold, and D.A. Iurino, "Semiaquatic Adaptations in a Giant Predatory Dinosaur," *Science* 2014, 345.6204: 1613–1616, doi: 10.1126/science. 1258750.

2 Ibid.

3 Ibid.

4 A.A. Snelling, "A Creationist's Best Friend," *Answers* 2014, 9.4:56–63.

5 Ibid.

6 S.G. Platt, R.M. Elsey, H. Liu, T.R. Rainwater, J.C. Nifong, A.E. Rosenblatt, M.R. Heithaus, and F.J. Mazzotti, "Frugivory and Seed Dispersal by Crocodilians: an Overlooked Form of Saurochory?" *Journal of Zoology* 2013, 291.2: 87–99, doi: 10.1111/jzo.12052.

7 F.E. Novas, L. Salgado, M. Suarez, F. Agnolin, M.D. Ezcurra, N.R. Chimento, R. de la Cruz, M.P. Isasi, A.O. Vargas, and D.Rubilar-Rogers, "An Enigmatic Plant-eating Theropod from the Late Jurassic Period of Chile," *Nature* 2015, doi: 10.1038/nature14307.

8 S. Lautenschlager, "Morphological and Functional Diversity in Therizinosaur Claws and the Implications for Theropod Claw Evolution," *Proceedings of the Royal Society B* 2014, 28.1785. 20140497. doi: 1098/rspb.2014.0497.

CHAPTER 6

1 M.P. Taylor and M.J. Wedel, "Why Sauropods Had Long Necks; and Why Giraffes Have Short Necks," *PeerJ* 2013, 1:e36.doi: 10.7717/peerj.36.

2 Ibid.

CHAPTER 8

1 J.B. Scannella and J.R. Horner, "*Torosaurus* Marsh, 1891, Is *Triceratops* Marsh, 1889 (Ceratopsidae: Chasmosaurinae): Synonymy through Ontogeny," *Journal of Vertebrate Paleontology* 2010, 30.4: 1157–1168.

2 M. Kaplan, "How to Eat a *Triceratops*," *Nature* 2012, 490.7421, doi:10.1038/nature.2012.11650

CHAPTER 10

1 G.M. Erickson, "On Dinosaur Growth," in *Annual Review of Earth and Planetary Sciences* 42, R. Jeanloz and K.H. Freeman, eds., Palo Alto, CA, 675–697, DOI: 10.1146/annurev-earth-060313-054858.

2 John R. Horner, personal communication.

3 V. Morell, "A Cold, Hard Look at Dinosaurs," *Discover* 1996, 17.12: 98–108.

4 J.A. Ruben, "The Metabolic Status of Some Late Cretaceous Dinosaurs," *Science* 1996, 273: 1204-1207.

5 L. Witmer, "The Evolution of the Antorbital Cavity of Archosaurs: A Study in Soft Tissue Reconstruction in the Fossil Record with an Analysis of the Function of Pneumaticity," *Journal of Vertebrate Paleontology* 17 1997 (Supplement to 1): 1–73.

6 J.A. Ruben, T.D. Jones, N.R. Geist, and W.J. Hillenius, "Lung Structure and Ventilation in Theropod Dinosaurs and Early Birds," *Science* 1997, 278.5341: 1267–1270.

7 J.A. Ruben, C. Dal Sasso, N.R. Geist, W.J. Hillenius, T.D. Jones, and M. Signore, "Pulmonary Function and Metabolic Physiology of Theropod Dinosaurs," *Science* 1999, 283.5401: 514–516.

8 R.A. Eagle, T. Tutken, T.S. Martin, A.K. Tripati, H.C. Fricke, M. Connely, R.L. Cifelli, and J.M. Eiler, "Dinosaur Body Temperatures Determined from Isotopic (^{13}C-^{18}O) Ordering in Fossil Biominerals," *Science* 2011, 333.6041: 443–445.

9 J.M. Grady, B.J. Enquist, E. Dettweiler-Robinson, N.A. Wright, and F.A. Smith, "Evidence for Mesothermy in Dinosaurs," *Science* 2014, 344.6189: 1268–1272.

10 G.M. Erickson, K.C. Rogers, and S.A. Yerby, "Dinosaurian Growth Patterns and Rapid Avian Growth Rates," *Nature* 2001, 412: 429–432; G.M. Erickson, P.J. Makovicky, P.J. Currie, M.A. Norell, S.A. Yerby, and C.A. Brochu, "Gigantism and Comparative Life-history Parameters of Tyrannosaurid Dinosaurs," *Nature* 2004, 430: 772–775.

11 T.L. Clarey, "Advocates for Cold-blooded Dinosaurs: The New Generation of Heretics," *GSA Today*. January 2007: 45-46. doi: 10.1130/GSAT01701GW.1.

12 Erickson, Rogers, and Yerby, "Dinosaurian Growth Patterns and Rapid Avian Growth Rates."

13 Erickson et al, "Gigantism and Comparative Life-history Parameters of Tyrannosaurid Dinosaurs," 772–775.

14 T.D. Jones, J.A Ruben, L.D. Martin, E.N. Kurochkin, A. Feduccia, P.F. Maderson, W.J. Hillenius, N.R. Geist, and V. Alifanov, "Nonavian Feathers in a Late Triassic Archosaur," *Science* 2000, 288.5474: 2202–2205.

15 A. Feduccia, T. Lingham-Soliar, and J.R. Hinchliffe, "Do Feathered Dinosaurs Exist? Testing the Hypothesis on Neontological and Paleontological Evidence," *Journal of Morphology* 2005, 266.2: 125–166.

16 S.A. Czerkas and A. Feduccia, "Jurassic Archosaur Is a Non-dinosaurian Bird," *Journal of Ornithology* 2014, 155.4: 841–851, DOI: 10.1007/s10336-014-1098-9.

17 Ruben et al, "Lung Structure and Ventilation in Theropod Dinosaurs and Early Birds."

18 P. Shipman, *Taking Wing: Archaeopteryx and the Evolution of Bird Flight* (New York, NY: Simon and Schuster, 1998).

19 A.M. Balanoff, G.S. Brever, T.B. Rowe, and M.A. Norell, "Evolutionary Origins of the Avian Brain," *Nature* 2013, 501: 93–96.DOI: 10.1038/nature12424.

20 C.P. Sloan, "Feathers for *T. rex*?" *National Geographic* 1999, 196.5: 98–107.

21 Balanoff et al, "Evolutionary Origins of the Avian Brain."

22 J. Wells, *Icons of Evolution: Science or Myth?: Why Much of What We Teach About Evolution Is Wrong* (Washington, DC: Regnery Publishing, 2000).

23 L.W. Simons, "*Archaeoraptor* Fossil Trail," *National Geographic* 2000, 198.4: 128–132.

24 M.H. Schweitzer, J.L. Wittmeyer, and J.R. Horner, "Gender-specific Reproductive Tissue in Ratites and *Tyrannosaurus rex*," *Science* 2005, 308.5727: 1456–1460.

25 A.H. Lee and S. Werning, "Sexual Maturity in Growing Dinosaurs Does Not Fit Reptilian Growth Models," *PNAS* 2008, 105.2: 582–587, Doi: 10.1073/pnas.0708903105.

26 Schweitzer et al, "Gender-specific Reproductive Tissue in Ratites and *Tyrannosaurus rex*."

27 P. R. Bell, N. E. Campione, W. S. Persons IV, P. J. Currie, P. L. Larson, D. H. Tanke, R. T. Bakker. "Tyrannosauroid integument reveals conflicting patterns of gigantism and feather evolution." Biology Letters. Published 7 June 2017.DOI: 10.1098/rsbl.2017.0092

28 T. Lingham-Soliar, "A Unique Cross Section Through the Skin of the Dinosaur *Psittacosaurus* from China Showing a Complex Fibre Architecture," *Proceedings of the Royal Society B* 2008, 275.1636: 775–780, doi: 10.1098/repb.2007.1342.

29 F. Zhang, S.L. Kearns, P.J. Orr, M.J. Benton, Z. Zhou, D. Johnson, X. Xu, and X. Wang, "Fossilized Melanosomes and the Color of Cretaceous Dinosaurs and Birds," *Nature* 2010, 463: 1075–1078, doi:10.1038/nature08740.

30 R.M. Carney, J. Vinther, M.D. Shawkey, L. D'Alba, and J. Ackermann, "New Evidence on the Colour and Nature of the Isolated *Archaeopteryx* Feather," *Nature Communications* 2011, 3.637, doi: 10.1038/ncomms1642.

31 Q. Li, K.Q. Gau, Q. Meng, J.A. Clarke, M.D. Shawkey, L. D'Alba, R. Pei, M. Ellison, M. Norell, and J. vinther, "Reconstruction of *Microraptor* and the Evolution of Iridescent Plumage," *Science* 2012, 335.6073: 1215, doi: 10.1126/science.1213780.

32 J. Lindgren, P. Sjovall, R.M. Carney, P. Uvdal, J.A. Gren, G. Dyke, B.P. Schultz, M.D. Shawkey, K.R. Barnes, and M.J. Polcyn, "Skin Pigmentation Provides Evidence of Convergent Melanism in Extinct Marine Reptiles," *Nature* 2014, 506: 484–488, doi: 10.1038/nature12899

33 D.B. Weishampel, "Acoustic Analysis of Potential Vocalization in Lambeosaurine Dinosaurs (Reptilia: Ornithischia)," *Paleobiology* 1981, 7.2: 252–261.

34 J.J.S. Johnson, "Sound Science about Dinosaurs," *Acts & Facts* 2015, 44.1: 18.

CHAPTER 11

1 K. Carpenter, *Eggs, Nests, and Baby Dinosaurs* (Bloomington and Indianapolis, IN: Indiana University Press, 1999).

2 J.R. Horner and J. Gorman, *Digging Dinosaurs* (New York: Workman Publishing, 1988).

3 Q. Meng, J. Liu, D.J. Varricchio, T. Huang, and C, Gao, "Parental Care in an Ornithischian Dinosaur," *Nature* 2004, 431: 145–146.

4 E.H. Colbert, *Men and Dinosaurs: The Search in Field and Laboratory* (New York: E.P. Dutton & Co., Inc., 1968).

5 N.R. Longrich, J.R. Horner, G.M. Erickson, and P.J. Currie, "Cannibalism in *Tyrannosaurus rex*," *PLoS ONE* 2010, 5.10: e13419. doi:10.1371/journal.pone.0013419.

6 S.L. Brusatte, *Dinosaur Paleobiology* (Chichester, West Sussex, England: John Wiley and Sons, Ltd., 2012), p. 169–171.

7 R.A. DePalma II, D.A. Burnham, L.D. Martin, B. Rothschild, and P.L. Larson, "Physical Evidence of Predatory Behavior in *Tyrannosaurus rex*." *PNAS Early Edition* 2013: 12560–12564, Doi: 10.1073/pnas.1216534110.

8 C. Carbone, S.T. Turvey, and J. Bielby, "Intra-guild Competition and Its Implications for One of the Biggest Terrestrial Predators, *Tyrannosaurus rex*," *Proc. R. Soc. B* 2011 278 1718 2682-2690doi:10.1098/rspb.2010.2497 published ahead of print January 26, 2011, 1471-295.

9 K.T. Bales and P.L. Falkingham, "Estimating Maximum Bite Performance in *Tyrannosaurus rex* Using Multi-body Dynamics," *Biology Letters*, 2012, Doi: 10.1098/rsbl.2012.0056.

10 W.I. Sellers and P.L. Manning, "Estimating Dinosaur Maximum Running Speeds Using Evolutionary Robotics," *Proc. R. Soc. B* 2007, 274: 2711-2716. doi: 10.1098/rspb.2007.0846.

11 Colbert, *Men and Dinosaurs: The Search in Field and Laboratory.*

12 Sellers and Manning, "Estimating Dinosaur Maximum Running Speeds Using Evolutionary Robotics."

13 R.T. McCrea, L.G. Buckley, J.O. Farlow, M.G. Lockley, P.J. Currie, N. A. Matthews, and S.G. Pemberton, "A 'Terror of Tyrannosaurs': The first Trackways of Tyrannosaurids and Evidence of Gregariousness and Pathology in Tyrannosauridae," *PLOS One* 2014, 9.7:13.e103613.

14 L.D. Xing, M.G. Lockley, J.P Zhang, A.R.C. Milner, H. Klein, D.Q. Li, W.S. Persons IV, and J. F. Ebi, "A New Early Cretaceous Dinosaur Track Assemblage and the First Definite Non-avian Theropod Swim Trackway from China," *Chinese Science Bulletin* 2013, 58.19: 2370–2378, doi: 10.1007/s11434-013-5802-6.

15 J. Morris, *The Global Flood: Unlocking Earth's Geologic History* (Dallas, TX: Institute for Creation Research, 2012); J. Morris and J.J.S. Johnson, "The Draining Floodwaters: Geologic Evidence Reflects the Genesis Text," *Acts & Facts* 2012, 41.1: 12–13; M.J. Oard, *Dinosaur Challenges and Mysteries* (Atlanta, GA: Creation Book Publishers, 2011); B. Thomas, "New Dinosaur Tracks Study Suggests Cataclysm. Creation Science Update," posted on www.icr.org January 25, 2013, accessed March 1, 2013.

16 A. Romilio, R.T. Tucker, and S.W. Salisbury. "Reevaluation of the Lark Quarry Dinosaur Tracksite (Late Albian–Cenomanian Winton Formation, Central-western Queensland, Australia): No Longer a Stampede?" *Journal of Vertebrate Paleontology* 2013, 33.1: 102–120.

17 D.J. Varricchio, P.C. Sereno, Z. Xijin, T. Lin, J.A. Wilson, and G.H. Lyon, "Mud-trapped Herd Captures Evidence of Distinctive Dinosaur Sociality," *Acta Palaeontologica Polonica* 2008, 53.4:567–578.

18 R.M. Alexander, "Estimates of Speeds of Dinosaurs," *Nature* 1976, 261:129–130.

19 M. Lockley, *Tracking Dinosaurs: A New Look at an Ancient World* (Cambridge, England: Cambridge University Press, 1991).

20 McCrea et al, "A 'Terror of Tyrannosaurs': The First Trackways of Tyrannosaurids and Evidence of Gregariousness and Pathology in Tyrannosauridae."

CHAPTER 12

1 L.W. Alvarez, W. Alvarez, F. Asaro, and H.V. Michel, "Extraterrestrial Cause for the Cretaceous-Tertiary Extinction," *Science* 1980, 208.4448: 1095–1108.

2 F.B. Bohor, E.E. Foord, P.J. Modreski, and D.M. Triplehorn, "Mineralogic Evidence for an Impact Event at the Cretaceous-Tertiary Boundary," *Science* 1984, 224.4651: 867–869.

3 D.B. Carlisle, "Diamonds at the K/T Boundary," *Nature* 1992, 357: 119–120.

4 A.R. Hildebrand, G.T. Penfield, D.A. Kring, M. Pilkington, A. Camargo Z., S.B. Jacobsen, and W.V. Boynton, "Chicxulub Crater: A Possible Cretaceous/Tertiary Boundary Impact Site on the Yucatan Peninsula, Mexico," *Geology* 1991, 19: 867–871.

5 D. Raup and J. Sepkoski, "Periodicity of Extinctions in the Geologic Past," *Proc. Natl. Acad. Sci. USA* 1984, 81: 801–805.

6 C. Officer and J. Page, *The Great Dinosaur Extinction Controversy* (Reading, MA: Helix Books, 1996).

7 Gerta Keller and T. Adatte, "Age and Paleoenvironment of Deccan Volcanism and the K-T Mass Extinction," *Geological Society of America Annual Meeting Abstracts with Programs* 2007, 39.6: 566.

8 F.S. Paquay, G.E. Ravizza, T.K. Dalai, and B. Peucker-Ehrenbrink, "Determining Chrondritic Impactor Size from the Marine Osmium Isotope Record," *Science* 2008, 320. 5873: 214-218.

9 B. Hayes, "Life Cycles," *American Scientist* 2005, 93.4: 299–303.

10 T.L. Clarey and K.A. Heim, "A Review of Extinction Patterns Across the K-Pg Boundary," *Journal of Creation Theology and Science Series C: Earth Sciences* 2012, 2:1–2.

11 J.R. Horner and J. Gorman, *Digging Dinosaurs* (New York, NY: Workman Publishing Company, 1988).

CHAPTER 14

1 R.B. J. Benson, N.E. Campione, M.T. Carrano, P.D. Mannion, C. Sullivan, P. Upchurch, and D.C. Evans, "Rates of Dinosaur Body Mass Evolution Indicate 170 Million Years of Sustained Ecological Innovation on the Avian Stem Lineage," *PLOS Biology* 2014, 12.5, e1001853.doi: 10.1371/journal.pbio.1001853.

2 Here the Bible is using the principle of accommodation — explaining God's feelings in the best way that people can understand. It wasn't that God literally "repented" in the sense of changing His mind, or regretting His previous actions. We know that God does not change (Malachi 3:6) or repent (Numbers 23:19). Rather, this word best describes God's feelings — His sadness and disappointment in humanity — in a way that we can understand.

3 S.A. Austin, J.R. Baumgardner, D.R. Humphreys, A.A. Snelling, L. Vardiman, and K.P. Wise, "Catastrophic Plate Tectonics: A Global Flood Model of Earth History," in *Proceedings of the third International Conference on Creationism, Technical Symposium Sessions*, R.E. Walsh, ed. (Pittsburgh, PA: Creation Science Fellowship, 1994), 609–/621.

4 A.A. Snelling, "Geophysical Issues: Understanding the Origin of the Continents, Their Rock Layers and Mountains," in *Grappling with the Chronology of the Genesis Flood*, S.W. Boyd and A.A. Snelling, eds. (Green Forest, AR: Master Books, 2014), 111–143.

5 Ibid.

PHOTO CREDITS:

T-top, B-bottom, L-left, R-right, C-Center

Answers in Genesis: pg 76-77

Bill Looney: pg 13 (all), pg 14 (B), pg 15 (B), pg 19 (T), pg 23 (B), pg 33, pg 37, pg 43 (T), pg 44 (BR), pg 54 (BR), pg 67 (BR), pg 73 (TR), pg 74, pg 81 (TL, BL), pg 82, pg 121, pg 150, pg 174 (T), pg 180

Bureau of Land Management (flickr.com/photos/mypubliclands/): pg 61 (all)

Christian Reinboth (flickr.com/photos/christian-reinboth/): pg 95

Creation Research Society Quarterly (2015, vol. 51, p. 255). Used by permission, pg 49 (B)

Dallas Krentzel (flickr.com/photos/31867959@N04/): pg 96

DncnH (flickr.com/photos/duncanh1/): pg 86 (TL)

Don DeBold (flickr.com/photos/ddebold/): pg 59

Dr. Tim Clarey: pg 15, (T), pg 19 (B), pg 36, pg 42 (BR), pg 117 (L), pg 124, pg 126 (B), pg 128 (all), pg 132, pg 135 (TL, BL), pg 142 (all), pg 151, pg 152 (all), pg 155 (all), pg 162, pg 167–171 (all), pg 174 (B), pg 176 (TR), pg 182, pg 183 (T)

dreamstime.com: pg 38 Afhunta, pg 109 (B) Mikekwok, pg 113 (B) Mengzhang, pg 119 (T) Pixelchaos

Eddy DR, Clarke JA (2011) New Information on the Cranial Anatomy of Acrocanthosaurus atokensis and Its Implications for the Phylogeny of Allosauroidea (Dinosauria: Theropoda). PLoS ONE 6(3): e17932. doi:10.1371/journal.pone.0017932; © 2011 Eddy, Clarke. This is an open-access article distributed under the terms of the Creative Commons Attribution License, which permits unrestricted use, distribution, and reproduction in any medium, provided the original author and source are credited. pg 130 (B)

Eden, Janine and Jim (flickr.com/photos/edenpictures/): pg 118 (MR)

Giovanni Dall'Orto, April 22 2007, pg 49 (T)

Harry Nguyen (flickr.com/photos/harrynguyen/): pg 125

James Emery (flickr.com/photos/emeryjl/): pg 70 (T)

Jeffrey Beall (flickr.com/photos/denverjeffrey/): pg 60 (T)

Jim Moore (flickr.com/photos/jdigger/): pg 80 (B)

Kabacchi (flickr.com/photos/kabacchi/): pg 23 (T)

Katie Thebeau (flickr.com/photos/katiethebeau/): pg 10

Leszek Leszczynski (flickr.com/photos/leszekleszczynski/); pg 115 (BR)

Liangtai Lin (flickr.com/photos/fossil_lin/): pg 122 (BL)

Marcus Winter (flickr.com/photos/ghazzog/): pg 94

Mike Bowler (https://www.flickr.com/photos/mbowler/): pg 71 (all)

NASA/JPL: pg 159 (ML, BL), pg 159 (T)

nps.gov: pg 31 (BL), pg 50

Peterson JE, Dischler C, Longrich NR (2013) Distributions of Cranial Pathologies Provide Evidence for Head-Butting in Dome-Headed Dinosaurs (Pachycephalosauridae). PLoS ONE 8(7): e68620. doi:10.1371/journal.pone.0068620; © 2013 Peterson et al. This is an open-access article distributed under the terms of the Creative Commons Attribution License, which permits unrestricted use, distribution, and reproduction in any medium, provided the original author and source are credited. pg 113 (TL)

Pieter VanderLinden (flickr.com/photos/pierelint/): pg 86 (ML)

Public Domain: pg 7 (B) Popular Science Monthly Volume 58, pg 17 Richard Owen, Memoirs on the extinct wingless birds of New Zealand. Vol. 2. London: John van Voorst, 1879, plate XCVII, pg 18 (TL), pg 21, pg 25 (T), pg 35 (BL), pg 35 (BM) Osborn, Henry Fairfield (1MR912). "Integument of the Iguanodont Dinosaur Trachodon" (PDF). Memoirs of the American Museum of Natural History v. 1: Plate VI., pg 52, pg 53 (MR), pg 53 (L), pg 53 (B), pg 60 (BL), pg 63 (all), pg 89 (B) Originally published in the Washington Post, Sep 24, 1924, in the article "Tail of Dinosaur Causes Flurry at Smithsonian." pg 122 (R) John Day Fossil Beds National Monument, pg 137 (all) Osborn, Henry Fairfield (1912). "Integument of the Iguanodont Dinosaur Trachodon" (PDF). Memoirs of the American Museum of Natural History v. 1: Plate VII + VII., pg 146 (T), pg 160, pg 164 (T), pg 165 (MR fossil) Anagoria

Raimond Spekking: pg 92-93, CC BY-SA 4.0 (via Wikimedia Commons)

Richo-Fan (flickr.com/photos/richo-fan/): pg 78

San Antonio JD, Schweitzer MH, Jensen ST, Kalluri R, Buckley M, Orgel JPRO (2011) Dinosaur Peptides Suggest Mechanisms of Protein Survival. PLoS ONE 6(6): e20381. doi:10.1371/journal.pone.0020381; © 2011 San Antonio et al. This is an open-access article distributed under the terms of the Creative Commons Attribution License, which permits unrestricted use, distribution, and reproduction in any medium, provided the original author and source are credited. pg 48 (all)

Science Photo Library: pg 27 Paul D Stewart/Science Photo Library, pg 28 (T) Gary Hincks / Science Source, pg 47 James King-Holmes / Science Source, pg 54 (TL, TR) Natural History Museum, London/Science Photo Library, pg 54 (TL) Natural History Museum, London/ Science Photo Library, pg 55 Paul D Stewart/Science Photo Library, pg 65 (BR), pg 65 (T) Natural History Museum, London/Science Photo Library, pg 83 (B) Joe Tucciarone/Science Photo Library, pg 103 Julius T Csotonyi/Science Photo Library, pg 105 Walter Myers/Science Photo Library, pg 106-107 (T) Christian Darkin/Science Photo Library, pg 123 (R) Leonello Calvetti/Science Photo Library, pg 135 (TM) Colin Keates (c) Dorling Kindersley, Courtesy of the Natural History Museum, London, pg 135 (BM) Natural History Museum, London/Science Photo Library, pg 136 (B) Carlos Goldin/Science Photo Library, pg 145 Mark Williamson/ Science Photo Library, pg. 147 (B all) Walter Myers/Science Photo Library, pg 148 David Woodfall Images/Science Photo Library, pg 153 David Nunuk/Science Photo Library, pg 154 David Nunuk/Science Photo Library, pg 159 (BR) (c) David Parker / Science Source

Shutterstock.com: pg 12 (C), pg 14 (T), pg 20 (B), pg 22 (B), pg 25 (B), pg 30, pg 31 (R), pg 32 (T), pg 34 (L), pg 35 (BR), pg 42 (BL), pg 43 (B), pg 44 (L), pg 54 (BL), pg 80 (C), pg 86 (BR), pg 90 (BL), pg 109 (T), pg 115 (T) (BL), pg 117 (BR), pg 139, pg 163, pg 165 (all), pg 172, pg 177 (T), pg 178, pg 181

Storem (flickr.com/photos/storem/); pg 75

Superstock.com: pg 24, pg 57 (B), pg 68, pg 83 (T), pg 85, pg 87, pg 88-89, pg 103 (T), pg 104 (TR), pg 111 (T), pg 114, pg 123 (R), pg 127

Tim Evanson: pg 6 (T, B), pg 34 (R), pg 99 (B), pg 141

Wikimedia Commons: pg 4, pg 5 Jordi Payà from Barcelona, Catalonia, pg 7 (T) © BrokenSphere / Wikimedia Commons., pg 8-9 H. Zell, pg 11 Greg from New York, NY, America, pg 12 (B) Drake411, pg 16 Kevin Walsh, pg 18 (TM, TR, BM, BR), pg 20 (T) Mike Peel (www.mikepeel.net), pg 32 (BL, BRM) Ghedoghedo, pg 32 (BLM) Kozuch, pg 32 (BR) Didier Descouens, pg 35 (T) Ghedoghedo, pg 39 Philip Maise, pg 44 (TR) Nobu Tamura, pg 45 (T) Nobu Tamura, pg 45 (M) Kumiko, Tokyo, Japan, pg 45 (B) Ryan Somma, pg 46 (BL) Gul-79, pg 56 (all), pg 58 J.M. Luijt, cc-by-2.5-nl licentie van auteur, pg 60 (BR) © Ad Meskens / Wikimedia Commons, pg 62 James St. John, pg 64-65 (B), pg 66 w:en:User:Stephan Schulz, pg 67 (T) GFDL, pg 67 (BL), pg 72 (BR) User:Bwark, pg 72 (BL) Zenhaus, pg 73 (BL) edmondo gnerre, pg 73 (BR) PePeEfe, pg 73 (TL) Eden, Janine and Jim from New York City, pg 79 ScottRobertAnselmo, pg 80 (T) Matt Martyniuk, pg 81 (R) Matt Martyniuk, pg 84 FunkMonk, pg 86 (BL), pg 90 (BR) Elika & Shannon, pg 91 (B) MCDinosaurhunter, pg 91 (T) Matt Wedel, pg 93 (T) ScottRobertAnselmo, pg 97 (all), pg 98 (B) Ghedoghedo, (T) Ballista, pg 100 missbossy, Singapore, pg 100 (T) Robin Zebrowski, pg 101 (B) Kabacchi, pg 102 Agsftw, pg 107 (B) Jean-Etienne Minh-Duy Poirrier from Bruxelles, Belgium, pg 108 Yuya Tamai from Gifu, Japan, pg 110 (all), pg 111 (B) http://www.app.pan.pl/article/item/app20120128.html, pg 112 Didier Descouens, pg 113 (TR, MR) John R. Horner, Mark B. Goodwin, pg 118 (T) Ryan Somma, pg 118 (BL) Shriram Rajagopalan from Vancouver, Canada, pg 119 (B) Christophe Hendrickx, pg 120 Tadek Kurpaski, pg 126 (T) Matt Martyniuk, pg 129 Eduard Solà, pg 131 Dinoguy2, pg 134 (B) Beatrice Murch, Buenos Aires, Argentina, pg 134 (T) http://www.ploscollections.org/article/info:doi/10.1371/journal.pone.0078733;jsessionid=441A913F8D576BBA46BF0960D01599FD, pg 135 (TR) Paxson Woelber, pg 136 (T) Didier Descouens, pg 143 Bruce McAdam, pg 144 Eduard Solà, pg 147 (T) D. Gordon E. Robertson, pg 149 Zenhaus, pg 156, pg 158 Nkansahrexford, pg 166 UNED Universidad Nacional de Educación a Distancia, pg 176 (B, MR, M, ML, TL), pg 183 (B) Khruner

Images from Wikimedia Commons are used under the CC-BY-SA-2.0, CC-BY-SA-3.0, CC-BY-SA-4.0 International license or the GNU Free Documentation License, Version 1.3.

Zachi Evenor (flickr.com/photos/zachievenor/): pg 104 (B)

INDEX: